PARTISAN FAMILIES

People decide about political parties by taking into account the preferences, values, expectations, and perceptions of their family, friends, colleagues, and neighbors. As most persons live with others, members of their households influence each other's political decisions. How and what they think about politics and what they do are the outcomes of social processes.

Applying varied statistical models to data from extensive German and British household surveys, this book shows that wives and husbands influence each other; young adults influence their parents, especially their mothers. Wives and mothers sit at the center of households: their partisanship influences the partisanship of everyone else, and the others affect them.

Politics in households interacts with competition among the political parties to sustain bounded partisanship. People ignore one of the major parties and vary their preference of its major rival over time. Election campaigns reinforce these choices.

Alan S. Zuckerman is a professor of political science at Brown University and a research professor at the DIW (German Institute of Economic Research). He has also served as a visitor at the University of Essex (British Academy Fellow); Tel-Aviv University (Fulbright Professor and Visiting Professor); the Hebrew University of Jerusalem (Lady Davis Visiting Professor); the University of Pisa (Fulbright Professor); the Istituto di Scienze Umane, Florence, Italy (Visiting Professor); and Stanford University (visiting associate). Professor Zuckerman has also published many books on diverse topics in political and social science and many articles in the leading journals of these fields.

Josip Dasović is an adjunct lecturer in political science at the University of Richmond. He is completing his Ph.D. in political science at Brown University. Prior to graduate school, he worked extensively in the former Yugoslavia with, among other organizations, the Croatian Helsinki Committee for Human Rights, at which he was a human rights activist and the organization's liaison to the International Criminal Court for the Former Yugoslavia (ICTY).

Jennifer Fitzgerald (Ph.D. Brown University, M.A. University of Chicago, B.A. Indiana University) is assistant professor of political science at the University of Colorado at Boulder. She has also been a visiting researcher at the DIW in Berlin, Germany, and the Centre d'étude et de recherche sur la vie locale (CERVL) in Bordeaux, France. Her research interests center around the influence of social context on political behavior and the social dimensions of immigration politics in Western Europe.

CAMBRIDGE STUDIES IN PUBLIC OPINION AND POLITICAL PSYCHOLOGY

Series Editors

Dennis Chong, *Northwestern University*
James H. Kuklinski, *University of Illinois, Urbana-Champaign*

Cambridge Studies in Public Opinion and Political Psychology publishes innovative research from a variety of theoretical and methodological perspectives on the mass public foundations of politics and society. Research in the series focuses on the origins and influence of mass opinion, the dynamics of information and deliberation, and the emotional, normative, and instrumental bases of political choice. In addition to examining psychological processes, the series explores the organization of groups, the association between individual and collective preferences, and the impact of institutions on beliefs and behavior.

Cambridge Studies in Public Opinion and Political Psychology is dedicated to furthering theoretical and empirical research on the relationship between the political system and the attitudes and actions of citizens.

Books in the series are listed on the page following the Index.

PARTISAN FAMILIES

The Social Logic of Bounded Partisanship in Germany and Britain

ALAN S. ZUCKERMAN
Brown University

JOSIP DASOVIĆ
University of Richmond

JENNIFER FITZGERALD
University of Colorado at Boulder

CAMBRIDGE UNIVERSITY PRESS
Cambridge, New York, Melbourne, Madrid, Cape Town, Singapore, São Paulo, Delhi

Cambridge University Press
32 Avenue of the Americas, New York, NY 10013-2473, USA

www.cambridge.org
Information on this title: www.cambridge.org/9780521874403

© Alan S. Zuckerman, Josip Dasović, Jennifer Fitzgerald 2007

This publication is in copyright. Subject to statutory exception
and to the provisions of relevant collective licensing agreements,
no reproduction of any part may take place without
the written permission of Cambridge University Press.

First published 2007

Printed in the United States of America

A catalog record for this publication is available from the British Library.

Library of Congress Cataloging in Publication Data

Zuckerman, Alan S., 1945–
Partisan families : the social logic of bounded partisanship in Germany and Britain /
Alan S. Zuckerman, Josip Dasović, Jennifer Fitzgerald.
  p. cm – (Cambridge studies in public opinion and political psychology)
Includes bibliographical references and index.
ISBN 978-0-521-87440-3 (hardback) – ISBN 978-0-521-69718-7 (pbk.)
1. Political participation – Social aspects – Germany. 2. Political participation – Social
aspects – Great Britain. 3. Political socialization – Germany. 4. Political socialization –
Great Britain. 5. Political sociology. I. Dasović Josip.
II. Fitzgerald, Jennifer, 1972– III. Title. IV. Series.
JN3971.A91Z828 2007
306.20941 – dc22   2006036503

ISBN 978-0-521-87440-3 hardback
ISBN 978-0-521-69718-7 paperback

Cambridge University Press has no responsibility for
the persistence or accuracy of URLs for external or
third-party Internet Web sites referred to in this publication
and does not guarantee that any content on such
Web sites is, or will remain, accurate or appropriate.

Contents

List of Figures and Tables	*page* ix
Acknowledgments	xiii
Preface: The Theoretical Approach, the Question, and the Data	xv

1.	The Social Logic of Partisanship: A Theoretical Excursion	1
2.	Bounded Partisanship in Germany and Britain	32
3.	A Multivariate Analysis of Partisan Support, Preference, and Constancy	47
4.	Bounded Partisanship in Intimate Social Units: Husbands, Wives, and Domestic Partners	71
5.	Bounded Partisanship in Intimate Social Units: Parents and Children	91
6.	Partisan Constancy and Partisan Families: Turnout and Vote Choice in Recent British Elections	123
	Conclusion: Family Ties, Bounded Partisanship, and Party Politics in Established Democracies	142

Appendix	161
References	167
Index	183

List of Figures and Tables

FIGURES

2.1	Aggregate Party Preferences in Germany	page 34
2.2	Aggregate Party Preferences among Germans in All Waves	35
2.3	Aggregate Party Preferences in Britain	36
2.4	Aggregate Party Preferences among Britons in All Waves	37
2.5	Aggregate Levels of Political Interest in Germany	39
2.6	Aggregate Levels of Political Interest in Britain	39
2.7	Partisan Constancy in Germany	43
2.8	Partisan Constancy in Britain	43
3.1	Household Influence on Partisan Choice in Germany	56
3.2	Household Influence on Partisan Choice in Britain	62
3.3	Household Influence on Partisan Constancy in Germany	67
3.4	Household Influence on Partisan Constancy in Britain	68
4.1	Reciprocal Effects of Partisan Choice in Germany	77
4.2	Reciprocal Effects of Partisan Choice in Britain	79
4.3	Joint Partisan Constancy in Germany	82
4.4	Joint Partisan Constancy in Britain	82
4.5	Postestimation Probabilities: Years of Marriage and Partisan Agreement in Germany	86
4.6	Postestimation Probabilities: Years of Marriage and Partisan Agreement in Britain	88
5.1	Trends in Aggregate Partisan Support in German Households	96
5.2	Trends in Aggregate Partisan Support in British Households	97
5.3	Trends in Partisan Support in German Households (among persons in all waves)	98

List of Figures and Tables

5.4	Trends in Partisan Support in British Households (among persons in all waves)	99
5.5	Bounded Partisan Choice among Young Persons in Germany	99
5.6	Bounded Partisan Choice among Young Persons in Britain	100
5.7	Postestimation Probabilities of Child's Partisan Support in Germany	105
5.8	Postestimation Probabilities of Child's SPD and CDU/CSU Choice	106
5.9	Postestimation Probabilities of Child's Partisan Support in Britain	109
5.10	Postestimation Probabilities of Child's Labour and Conservative Choice	109
5.11	Political Influence in Households	121
6.1	Parental Influence on Child's Labour Vote	133
6.2	Influence of Husband and Child on Mother's Labour Vote	134
6.3	Influence of Wife and Child on Father's Labour Vote	134
6.4	Parental Influence on Child's Conservative Vote	137
6.5	Influence of Husband and Child on Mother's Conservative Vote	137
6.6	Influence of Wife and Child on Father's Conservative Vote	138

TABLES

2.1	Partisan Change between any Two Adjacent Years	41
2.2	The Electorates Distributed by Partisan Choice between Two Points in Time	42
3.1	Partisan Choice for SPD – Heckman Probit Selection Model	52
3.2	Partisan Choice for CDU/CSU – Heckman Probit Selection Model	54
3.3	Partisan Choice for Labour – Heckman Probit Selection Model	58
3.4	Partisan Choice for Conservatives – Heckman Probit Selection Model	60
3.5	Partisan Choice and Constancy in Germany (1985–2001) – Zero-Inflated Negative Binomial Regression Model	64
3.6	Partisan Choice and Constancy in Britain (1991–2001) – Zero-Inflated Negative Binomial Regression Model	66
4.1	Reciprocal Effects on SPD Choice – Instrumental Variable Probit Model	75
4.2	Reciprocal Effects on CDU/CSU Choice – Instrumental Variable Probit Model	76

List of Figures and Tables

4.3	Reciprocal Effects on Labour Choice – Instrumental Variable Probit Model	78
4.4	Reciprocal Effects on Conservative Choice – Instrumental Variable Probit Model	79
4.5	Partisan Agreement between Husband and Wife for the SPD – Time-Series Logit Model	84
4.6	Partisan Agreement between Husband and Wife for the CDU/CSU – Time-Series Logit Model	85
4.7	Partisan Agreement between Husband and Wife for Labour – Time-Series Logit Model	87
4.8	Partisan Agreement between Husband and Wife for Conservatives – Time-Series Logit Model	87
5.1	Agreement among Mothers, Fathers, and Children in Germany – Polychoric Correlation Coefficients	93
5.2	Agreement among Mothers, Fathers, and Children in Britain – Polychoric Correlation Coefficients	93
5.3	Partisan Choice across Generations in German and British Households	95
5.4	Choice of SPD among Young Persons in Germany – Heckman Probit Selection Model	103
5.5	Choice of CDU/CSU among Young Persons in Germany – Heckman Probit Selection Model	104
5.6	Choice of Labour among Young Persons in Britain – Heckman Probit Selection Model	107
5.7	Choice of the Conservatives among Young Persons in Britain – Heckman Probit Selection Model	108
5.8	Agreement on SPD Choice in Germany among Mothers, Fathers, and Children – Three-Stage Regression Model	112
5.9	Agreement on CDU/CSU Choice in Germany among Mothers, Fathers, and Children – Three-Stage Regression Model	114
5.10	Agreement on Labour Choice in Britain among Mothers, Fathers, and Children – Three-Stage Regression Model	117
5.11	Agreement on Conservative Choice in Britain among Mothers, Fathers, and Children – Three-Stage Regression Model	118
5.12	Predicted Probabilities of Partisan Choice	119
6.1	Labour Vote – Heckman Probit Selection Model	128
6.2	Conservative Vote – Heckman Probit Selection Model	129
6.3	Family Effects on Labor Vote for Mothers, Fathers, and Children – Three-Stage Regression Model	132

List of Figures and Tables

6.4 Family Effects on Conservative Vote for Mothers, Fathers, and Children – Three-Stage Regression Model   136
6.5 Participation in British General Elections – Logit Model   140
6.6 Frequency of Participation in British Local Elections – Ordered Logit Model   140

Acknowledgments

Many persons assisted our efforts to complete this research. It is a pleasure and a privilege to be able to thank them publicly.

Alan S. Zuckerman is grateful to DIW (German Institute for Economic Research) for an appointment as Research Professor, which facilitated two research trips to Berlin and fostered his collaboration on a related project with Martin Kroh. These visits enabled him to more deeply understand the tortured history and the routine present that make up Germany. Jennifer Fitzgerald too benefited from a research stay at DIW, and she would like to thank Deborah Bowen, Anita Drever, Bettina Isengard, Gundi Kniess, and Thorsten Schneider. We also want to thank other colleagues at DIW who are involved with the German Socio-Economic Panel Study (GSOEP), especially Gert Wagner, the director, Christina Kurka, the administrator, and Martin Kroh, a first-rate political scientist.

Zuckerman also takes pleasure in thanking the British Academy for a research fellowship that allowed him to work at the Institute for Social and Economic Research at the University Essex. This grant enabled him to benefit from Malcolm Brynin's deep familiarity with the British Household Panel Survey (BHPS). For assistance with the British data, we are also pleased to thank Jonathan Gershuny, director, and John Brice, who always answered our long-distance and sometimes panicky messages.

Our colleagues at GSOEP and BHPS bear no responsibility for how we have used the data that they have kindly made available to us.

Readers of this book encounter many numbers, models, and graphs. When these presentations succeed, it's because of the help that we received from Scott Allard, Andrew Foster, Rachel Friedberg, Joseph Harkness, James Heckman, Michael Herron, Joseph Hogan, Gary King, Martin Kroh, Scott Long, Michael White, and Ezra Zuckerman. We are honored to be able to thank them. When our analyses do not measure up, it's our fault.

Acknowledgments

This book began as a graduate seminar in the department of political science, Brown University. Over time, including many long days peering together at the computer in Zuckerman's office, we developed into a team. Each of our contributions was necessary to turn the hope of successful research into a book in which we take delight and pride. During that time too, we benefited from the assistance of our colleagues and friends in the department. In addition to those already mentioned, we are very pleased to thank Roger Cobb, Patti Gardner, and Wendy Schiller.

We are also very happy to thank Lew Bateman, our editor at Cambridge University Press, for his efficiency and encouragement, as well as Sara Black, our copy editor, Ernie Haim, production editor, and Shelby Peak, associate production controller, for their help in bringing our work to light.

Portions of Chapter 1 were published in Alan S. Zuckerman, ed., *The Social Logic Politics* by Temple University Press; they are reprinted here by permission of Temple University Press.

Alan Zuckerman wants also to thank his wife Ricki, for being Ricki, and to dedicate this and so much more to her.

Josip Dasović dedicates this book to his parents, Marko and Dragica Dasović, and also to Ivana.

Jennifer Fitzgerald offers this as a small token of appreciation to her amazing parents, John and Fay.

Preface: The Theoretical Approach, the Question, and the Data

> Social relationship: The behavior of a plurality of actors insofar as, in its meaningful content, the action of each takes account of that of the others and is oriented in these terms.
>
> Max Weber, *Economy and Society* I, 26

> To those [guards] who do not talk to you... you dare not speak. If you are fortunate enough to have someone next to you with whom you have a common language, good for you, you'll be able to exchange your impressions, seek counsel, let off steam, confide in him; if you don't find anyone, your tongue dries up in a few days, and your thought with it.
>
> Primo Levi, *The Drowned and the Saved*, 71

We offer a simple story. How do people decide about political parties? Much as they make other decisions. They take into account the preferences, values, expectations, and perceptions of their family, friends, colleagues, and neighbors.[1,2] People affect one another, and so any one decision responds to the particular mix of views in a person's social networks. As Max Weber, a founder of social science, taught, and as Primo Levi witnessed in the Holocaust: people live and experience their lives and their thoughts in social relationships. How and what they think about politics and what they do are the outcomes of social processes.

---

1 The work of the Columbia School of electoral sociology provides the classic literature for contemporary social science; see Berelson, Lazarsfeld, and McPhee (1954) and Lazarsfeld, Berelson, and Gaudet ([1948] 1968). Recent work includes Huckfeldt and Sprague (1995); Huckfeldt, Johnson, and Sprague (2004; 2005); Kenny (1994); Straits (1990; 1991); and the essays in Zuckerman (2005a). We develop this theme in the first chapter.
2 Berns et al. (2005); King-Cassas et al. (2005); Knickmeyer et al. (2002); Riling et al. (2002) present evidence of a neurological basis for social relationships as determinants of human behavior.

Preface

These simple claims entail complex analyses. Political influence, like other outcomes of social relationships, is usually reciprocal. Absent pure dominance of one member of a dyad over another, the logic that implies that A influences B also maintains that B affects A, and so on for each of the additional dyads in the social network. In order to detail the patterns of political influence, evidence needs to be applied to appropriate analytical models. Furthermore, political influence is probabilistic, not determined; people retain the possibility to go their own ways, and these relationships must be appropriately modeled as well. As we seek to account for partisan choices, we will specify the relative strength and causal flows among the relationships that we observe.

As most persons live with others, members of their households – usually husbands, wives, parents, and children – influence each other's political decisions.[3] This general statement directly implies critical points for our analysis. Frequency of interaction affects the probability of influence. Family members affect each other, in part at least, because they see each other frequently and so they send and receive many cues. Wives and husbands influence each other; it is wrong to assume that, in principle at least, one partner dominates the other. Similarly, children, especially those who are adults, influence their parents; the flow of influence need not extend solely from the older to the younger generation. By implication, therefore, political interest is not the source of influence within families. The results of our empirical analyses place wives/mothers at the center of households: their partisanship influences the partisanship of everyone else, and the others affect them; neither statement applies to husbands/fathers or children.

What do people decide about political parties?[4] They decide whether or not to support any political party (partisan support), which party to name (partisan preference or choice), and they make these decisions again and again over time (support and preference constancy). Most people think about a party as an object of support or preference, and most consider whether or not to vote for its candidates; few take part in the party's

---

[3] Classic works that examine political preferences between household partners include Dogan (1967); Glaser (1959–60); March (1953–4). See also De Graaf and Heath (1992); Kingston and Finkel (1987); Niemi, Hedges, and Jennings (1977); Stoker and Jennings (1995; 2005); Zuckerman and Kotler-Berkowitz (1998); Zuckerman, Fitzgerald, and Dasović (2005). There is a more substantial literature on political socialization within families; see for example Achen (2002); Beck and Jennings (1975; 1991); Davies (1970); Jennings and Niemi (1968); Niemi and Jennings (1991); Tedin (1974); Ventura (2001), and see Chapter 5. Recent research extends these claims to matters of happiness (Powdthavee 2005) and health (Wilson 2002; see Chapter 4).

[4] Later in the preface and in Chapter 2, we define the concepts associated with partisanship.

xvi

activities. In established democracies, partisanship and electoral behavior encompass much of the way that most people relate to politics. Even as people consider the political parties and cast ballots in elections, politics is not one of their daily activities. Taken in the aggregate, however, these decisions carry enormous power, conditioning the behavior of government leaders.

What kind of choice is the preference for a political party? All choices are not the same. Some are perceived to entail few if any consequences for the actor; some are not so understood. On this account, philosophers distinguish "picking," which applies to "small" decisions, from "choosing," which entails reasons and preferences (see Ullmann-Margalit and Morgenbesser 1977; Ullmann-Margalit 2005). The consequences of the choice may be distinguished with regard to the effect on the actor. Some choices are "big"; they are meant to change lives and may be classified as "opting" or "converting," where "drifting" implies that the decision will not transform the individual (Ullmann-Margalit 2005). Closer to a small than a big one, partisan choice is a preference; it is a decision for a reason.

When people choose a political party, are they seeking to advance an interest? In this study, we apply the principles of bounded rationality, not optimization, to partisan choice. In chess, for example, players seek to win; that is their interest. This "macroscopic driving force" (Aumann 2005) affects decisions about strategy and tactics. Maximizing expected utility guides economic choices, where "making money" serves as a shorthand for each person's interest. Similarly, "controlling power" may guide the analysis of the decisions of politicians. In all these cases, people decide how best to advance their interest by calculating with regard to clearly defined rules and the strategic choices of opponents. When a person announces a partisan preference or casts a ballot, there is no immediate and direct gain or loss. Selecting a party does not affect one's ability to advance an interest the way that capturing an opponent's queen, getting the best deal on a purchase, or landing a cabinet office does. Blais (2000) details the limited value of interest-based rational choice analyses of voting and by implication partisanship. Partisan preference involves choices about distant and abstract objects. The hope of somehow influencing government policies that might affect the actor's life does not provide the driving force for these decisions. Instead, we apply the principles of social learning to the effort to make the best possible choice in an arena without clearly defined interests.

At any given election, partisans usually vote for their party's candidates. This relationship is an empirical regularity, not a tautology. First, different survey questions define these concepts: one asks for self-reflection and another describes behavior, and so they differ in practice as well as principle. In the same survey, these responses may overlap. Asked again and

Preface

again over time, as in the surveys that we examine, temporal and analytical distance separates the answers about the political parties and reports of voting decisions. Equally as important, we show that the constancy of partisan choice varies, and this variation influences electoral decisions. Partisanship and electoral behavior are distinct clusters of concepts.

We detail the elements of partisanship, providing evidence of choices taken by persons in large longitudinal sample surveys in Germany (1985–2001) and Britain (1991–2001). Examining partisanship in only one year without regard to other years provides a limited and, we think, distorted picture of partisan behavior. Our German and British panel data enable us to address the more complex issue of choices made at different points in time.

The citizens of Germany and Great Britain are *bounded partisans*. Over time, most of them effectively turn their back on one of the major parties by never choosing to support it, and they vary their choice of the other large party – sometimes choosing it and sometimes not. They are generally more constant with regard to the party that they do not name than with their preferred party. Also, the responses of very few persons take them from party to party. People construct a choice set from the full list of political parties, eliminating one of the major parties and establishing the other as a possible object of support.

When we claim to explain partisanship, what do we mean? How do we account for the elements of partisanship and electoral behavior? First, drawing on the social logic of politics, we propose a set of explanatory mechanisms that emphasize the centrality of family and household interactions in partisan decisions. In the first chapter, we provide the intellectual history of these social mechanisms by examining the work of anthropologists, economists, decision theorists, social psychologists, and sociologists, as well as political scientists. Where appropriate and possible, we contrast our perspective with the expectations of other more widely used approaches: the claim of classical rational choice that people support political parties that are in line with their personal interests; the view that partisan choice is a form of social identification in which people develop psychological attachments to political parties; and the hypothesis that partisan support derives from attachments to a social class or ethnic group and the political party with which it is associated. We also test our claims against the processes and events that characterize Germany and Britain, our two cases for empirical analysis. We control for the effects of a secular decline in partisanship, which has characterized European electorates over the past two decades, and for particular events, such as recurrent elections, German Reunification, and the rise of Labour to power in Britain. We explain by justifying and testing a theory of the social logic of partisanship.

Preface

Our empirical analyses that follow revolve around probabilities. They specify the presence and strength of the relationships between predictor or explanatory variables and outcome or dependent variables. They compare the results to the mean probability in the sample for each of the dependent variables in our different models: partisan support, choice, constancy, turnout, and vote choice. They examine these in the context of the absolute values as well: the probability of naming or not naming a party, of agreeing with someone else's partisan choice, and so forth. We detail the impact of others in the household, net of the effects of other predictor variables, on the probability that a person is above or below the mean probability of the particular outcome.

As a result, a quick glance ahead to the pages of our book finds many tables and graphs, as we structure the empirical analysis around the results of statistical models. Naming a particular party rests on whether or not a person supports any party. Similarly, vote choice rests on the decision to go to the polls, turnout. Some of our models, therefore, analyze together partisan support and choice or join turnout and vote choice (Heckman Probit Selection models) and some link partisan choice and constancy (Zero-Inflated Negative Binomial models). When we examine the reciprocal relationships within households, we use instrumental probit and three-stage linear probability models for systems of simultaneous equations. These use instrumental variables and two-stage models, where we examine household partners, and three-stage models, where we focus on wives/mothers, husbands/fathers, and children. In turn, these models require that we examine one dependent variable at a time, and so we look only at the party named or vote choice, not partisan support or turnout. Each model has significant strengths and some weaknesses. Taken together with our theory, they offer a compelling account of the social logic of partisanship. And so, sometimes, we analyze with words, sometimes with numbers presented in tables, and sometimes with figures and graphs that help to interpret the statistical analyses.

We examine two established democracies, Germany and Britain. No matter Germany's checkered, brutal, and tragic political history, by the middle of the 1980s, four decades after the end of the Second World War, Germans had become accustomed to democratic rule. Happily, little remains of their past to challenge the claim that on the dimensions of our analysis their political system is not much different from that of the British. Indeed, we show again and again that at least with regard to partisanship Germany and Britain are very similar.

Overflowing with data on family members but absent information on anyone else, we apply the social logic of politics to political decisions made by members of German and British households. This focus responds to the data that we have; it does not imply that only family members

Preface

matter with regard to political decisions, but it does suggest that they do affect each other. We use survey responses to access both partisanship and electoral behavior, so for each concept we, like other scholars of mass political behavior, rely on what people say they think and do. Because of the importance of these surveys, the German Socio-Economic Panel Study (GSOEP) and the British Household Panel Survey (BHPS), we describe them here in the book's Preface.

OUR DATA

Like almost all other studies of partisanship and voting, our evidence comes from nationally representative surveys. Several characteristics distinguish the ones that we use. Each covers many adjacent years, offering an extensive panel that provides information on many more years than any other survey with information on partisanship. These years are not restricted to periods with elections, and so our analysis of partisanship is not limited to moments of competition among the parties. Focusing on German and British households, these surveys also interview all persons over the age of 15 in a household. The surveys do not rely on reports of only one member of the household; they also do not ask respondents to remember the partisan preferences of their parents. Because all family members usually live in the same household, we use the terms interchangeably. GSOEP and BHPS enable us to explore the reciprocal effects of family members on each other as they decide about partisan support and choice at a single point in time and over time, and as they decide about casting ballots in elections.

In our estimation, these are the best surveys available for the study of partisanship. Even so, they have limitations. Designed primarily for the use of labor economists, demographers, and scholars of community health, they do not contain the extensive battery of political questions that one finds in election surveys. Even as the data allow us to elaborate a social logic of politics, they do not directly permit us to confront other theoretical perspectives within political science.

There are also some technical considerations. These are unbalanced panel surveys, meaning that some persons respond in many, or even all, of the years, and some answer only once. Second, the respondents are usually persons who live together, and so we have a priori reason to expect their answers to resemble each other. We address these concerns in several ways. First, where appropriate, we apply controls that produce robust standard errors, thereby accounting for the interdependence of observations within clustered responses. Some of our models control directly for autocorrelation, occasioned by time-series data. In addition, the count models allow us to pay particular attention to persons who respond in all

Preface

waves. Because these people are most likely to remain in the most stable social relationships, we expect them to display particularly high rates of partisan constancy. Here, we purposefully study a subset of the population whose responses one might expect to be systematically different from the full sample. Finally and most obviously, the reciprocal models do not seek to erase the effect of household influences on partisanship and voting. To the contrary, they seek to specify these relationships. That, after all, is the point of the book.

GERMAN SOCIO-ECONOMIC PANEL STUDY

The German Socio-Economic Panel Study[5] interviews a large and representative sample of the German population. Beginning in 1984 with West Germans and immigrants and adding the East German sample in the months just before Reunification, it continues its interviews each year into the foreseeable future. We look at the years 1985–2001.[6] No other panel survey encompasses so many waves or so extensive a period of time; all others contain much smaller samples, and GSOEP provides a national – not just local – study of family networks. It is an exemplary panel survey, also because it follows persons who move into new households and then asks the battery of questions of all adults present there as well.[7] Of particular importance for our analysis is the fact that it interviews everyone present in the household over the age of 15, without relying on the reports of only one member, such as the head of household, as do many household surveys: 12,031 persons are interviewed at least once; the mean number of years for each person in the survey is 9.6. Combining the number surveyed each year with the number of waves produces 115,372 person-years. In turn, 2,997 respond in each and every wave. Because of the large sample, it is also possible to aggregate the responses into descriptions of the German states, the *Bundesländer*. We study Germans living in what was West Germany at the initiation of the survey (Zuckerman and Kroh 2005 extend the analysis to East Germans and immigrants, and see also Schmitt-Beck, Wieck, and Christoph 2006). Examining these Germans (and not all those in GSOEP's samples) enables us to examine partisanship over the

---

[5] Full descriptions of the survey may be obtained from the Web site of the Deutsche Institut für Wirtschaftsforschung (DIW Berlin): http://www.diw.de/GSOEP.
[6] Because of instability in the responses to the partisanship questions in the first year of our data set, our analysis begins in the second year, 1985.
[7] The survey follows respondents who move, but it is less able to maintain contact with those who are in temporary housing and people who move frequently (Kroh and Spiess 2004). As a result, even GSOEP probably understates the level of instability in the responses to questions on partisanship.

Preface

greatest number of years and provides a sample that is most comparable to the respondents in BHPS. GSOEP offers an extensive and detailed array of data for the exploration of micropartisanship, as well as many other elements of German social, economic, and political life.

Consider how GSOEP defines and measures partisanship. The English-language translation of the relevant questions reads: "Many people in the Federal Republic of Germany [Germany, after 1990] are inclined to a certain political party, although from time to time they vote for another political party. What about you: Are you inclined – generally speaking – to a particular party?" Those who respond "Yes" – we define as party supporters. They are then asked, "Which one?" and handed a card that lists all parties with seats in the Bundestag. This defines party preference (or choice), and we focus almost all of our attention on those who name the two dominant parties, the Social Democrats (SPD) and the Christian Democrats or the Christian Socials in Bavaria (CDU/CSU).[8]

The protocol repeats the set of questions in each and every one of GSOEP's waves. Because the opening question names no parties, it avoids problems of instrumentation that are associated with the traditional measure.[9] It does not presume an answer by offering the names of the parties

---

8 This question closely resembles the one used in the German national election and other political surveys, and the marginal results match these data as well (see Falter, Schoen, and Caballero 2000; Green, Palmquist, and Schickler 2002, 164–203; Norpoth 1984; Schickler and Green 1997, 463; Zelle 1998, 70). Because few respondents name the smaller parties, the Free Democrats, Greens, and the Party of Democratic Socialism, and because including these parties would add considerable complexity without significant analytical gain, we focus our attention on the two dominant parties.

GSOEP also asks a question about the strength of this inclination, which parallels the literature's concern with the strength of party identification. Examining the responses over time indicates that people offer inconsistent answers – moving among the categories of strength nonsystematically. This appears to be an ambiguous and unreliable question. Students of partisanship in the United States combine a question on party identification with one on the strength of that identification into a seven-point scale, which varies from strong Democrat at one extreme to strong Republican at the other. Because of the presence of the Free Democrats and Greens in Germany and the Liberal Democrats and the nationalist parties in Britain, the absence of much cross-party movement in Germany and Britain, and because of the unreliability of the strength measure, we do not use this scale.

9 Most versions of the traditional measure contain wording like the following: "Generally speaking do you think of yourself as an X, Y, or Z?" where the letters indicate the names of particular political parties. This question implies identification and contributes a specific answer to the question, thereby prompting a response. Presented again and again in a panel survey, it increases both the probability of an answer in each year and the same answer over time. This helps to account for the high levels of partisan stability found in Green, Palmquist, and Schickler (2002).

Preface

to those who would not otherwise be able to do so. Because it only asks the respondents to describe themselves, it does not prejudge the issue of psychological attachment. After all, one may support, prefer, or incline towards a party without identifying with it. Because GSOEP regularly taps party choices during and between electoral periods, political campaigns do not induce most of the responses, as they might in election surveys. These questions offer reliable and internally valid measures of partisan support in Germany.

The survey offers unparalleled opportunities to study the social logic of partisanship. By providing answers offered by each person in the household, the data enable us to focus on the social unit characterized by relatively high levels of interaction, trust and dependence, shared information and values, and the unit in which political discussion – both direct and verbal and indirect and nonverbal – is most likely to occur. Because family members vary in political interest, we can also see whether or not these asymmetries underpin variations in political influence within households. Offering information on the distribution of partisanship in the *Bundesländer* in which each respondent resides, GSOEP presents a measure of political context. In addition, questions about voluntary organizations, social contacts, and trade union and other social memberships tap social ties beyond the household. There are several indicators of social class (household income; education; occupation graded by Goldthorpe measures[10]); and a direct question about religious self-identification and attendance. Furthermore, there are assessments of worries about the economy. These data allow us to describe and model the partisanship of Germans.

Obviously, partisan choices occur in general political contexts as well as immediate social circumstances. Indeed, one might think that partisan preference directly responds to political events – although there is no evidence in our data to support this conjecture. Some of these political happenings are on-going, such as the declines in partisanship that began in the early 1970s and continue to apply to Germans and many other Europeans. Some are recurrent, such as national and local elections, whose campaigns remind persons of the need to select a party. Some are episodic, and in Germany these events give one reason to expect fundamental changes in partisanship.

The Germans who respond to GSOEP's questions lived through years of political stasis, then the sudden and rapid transformation of their state,

---

10 This measure of objective social class (occupation) specifies multiple locations, not a simple manual–nonmanual divide. Evans (1999b) elaborates the utility of this measure of social class in electoral and other analyses; see also Goldthorpe (1999a; 1999b).

Preface

and after that a change of governing coalition. German Reunification occurred in the fall of 1991, after two years of political turmoil. While the first five years of the survey cover a time that might be described as normal politics in an established democracy, subsequent years are anything but usual. The German Democratic Republic collapsed. Its regions became part of the German Federal Republic, and West Germany became Germany. Led by Chancellor Kohl and his Christian Democratic Union/Christian Social Union, the government made far-reaching decisions: East Germans immediately became citizens of the state, and their currency was made the equivalent of the German mark. Massive infusions of economic aid moved to the new areas of reunified Germany. By 1998, the long-governing chancellor, Helmut Kohl, lost the national election to Gerhard Schroeder and the Social Democrats, and the CDU/CSU moved to the opposition. A few years later, the euro replaced the mark as the official currency when Germany and its neighbors moved more deeply into the European Union. These political events provide conflicting cues. Whereas secular declines in partisanship indicate that citizens are distancing themselves from the parties, election campaigns focus attention on the need to choose. In turn, the transformation of the German state implies diverse factors, some of which lead to preference for one or the other of the parties and some do not, even as they would seem to rouse the attention of German citizens and their political interest. There is reason to expect variations in partisan choices to respond to these political contexts. In turn, we use these events and processes as controls for the demonstration of household effects of political influence.

BRITISH HOUSEHOLD PANEL SURVEY

Modeled on GSOEP, the British Household Panel Survey (BPS)[11] also surveys large numbers of people over many years, providing an on-going representative sample of Britons. We explore eleven waves of BHPS, 1991–2001, wherein 20,235 are interviewed at least once, with a mean number of interviews of five. Multiplying the number of persons by the number of years produces a data base of 100,533 person-years. In all, 4,989 Britons are interviewed in each and every wave. These data offer material for a fine-grained portrait of British social, economic, and political life, particularly as they play out within households over many years.

Although BHPS's three questions on partisanship differ slightly from the German survey, they too offer a measure with internal validity. Here,

---

[11] Full descriptions may be obtained from the Web site of the survey's home institution, the Institute for Social and Economic Research at the University of Essex: http://www.iser.essex.ac.uk.

Preface

the questions are: "Generally speaking do you think of yourself as a supporter of any one political party?" If the answer is "No," the survey then asks, "Do you think of yourself as a little closer to one political party than to the others?" We define those who say "Yes" to either question as party supporters. They are then asked, "Which one?" and volunteer the party's name (defining partisan choice or preference). Again, we study persons who name the major parties, Labour and the Conservatives (Tories).[12] Like their counterparts in GSOEP, these questions are asked in each and every wave. They too do not name the parties in the opening question, nor do they imply any kind of psychological identification. Like GSOEP's questions, this is a measure appropriate to the study of partisanship.[13]

BHPS, but not GSOEP, also includes questions on voting, and in Chapter 6 we analyze turnout and vote choice in Britain. These questions are straightforward. BHPS asks the respondents whether or not they voted in the general election of that year, and, if so, for the candidate of which party.

The questions also permit us to assess various theoretical perspectives on partisanship and electoral behavior in Britain. They provide measures of age, social class (subjective identification, Goldthorpe measures of occupation, and education), religion (identification and attendance at services), economic perceptions and concerns, membership in trade unions and various social organizations, and political interest. Here too we assess the impact of intimate social ties on partisan choice and persistence, and the large samples allow us to aggregate responses into pictures of Britain's regions.

The political context of partisan choice in Britain differs dramatically from that of Germany. Even so and again, we do not find evidence that particular political events influence partisan preferences. Nothing transforms the British state; there is no massive infusion of new citizens; the government does not provide extraordinary resources to a particular region of the country, and though public discussion swirls about the topic, the euro does not replace the British pound as the official currency. There is a point of similarity. In Britain the large party of the political left replaced a

---

12 Again too, we do not analyze partisanship with regard to the smaller parties, the Liberal Democrats, the Scottish Nationalists, and the Plaid Cymru, and also omit the strength of partisanship; we provide the reasons in footnote 6 in this Preface. We return to the matter in the Conclusion.
13 In a series of papers, Bartle (1999; 2001; 2003) offers a telling critique of the traditional measure, as it has been used in the British Election Studies. Blais et al. (2001) and Greene (2002) also offer trenchant critiques of the traditional question as an indicator of partisanship in the United States.

Preface

long-serving conservative government, as Tony Blair and the Labour Party removed John Major from the prime minister's office and turned the Conservative Party into an ineffective opposition. The BHPS data enable us to examine the relationship between the macroevents and the microchoices about the political parties during these years.

Both GSOEP and BHPS offer measures appropriate to each polity, and the differences between the two sets of questions do not deny the claim that each and both present indicators of the partisanship concept. Experts in electoral politics in Germany devised a measure that omits the names of the parties in the initial question, uses the verb phrase "incline towards," and then offers a list of the parties only to those who claim to support one. British scholars tap partisanship by using the verbs "support" and "close to" and never list the parties. Each is appropriate to the particular context. Together, they provide contextually proper measures of the same concept: partisanship.

These exceptionally useful national surveys (even unique, if it makes sense to claim two unique instances of anything) cover many more years and many more people than other surveys that contain information on partisanship. Because each begins with and maintains a very large sample, aggregating the responses allows us to control for the effects of particular events – national elections for one example, the fall of the Berlin wall for another – on partisan decisions. Similarly, we can observe and test for the presence of a secular decline in partisanship in both countries during these decades. Because these are panel surveys, we can explore partisan constancy and switching, not just choice at a single point in time. This enables us to see whether or not people display the persistent behavior that would characterize those who identify with a political party or the movement across the parties that would be expected of partisan consumers. Looking at the surveys as an aggregated cross section and at the panel results also enables us to disentangle the causal processes that relate to time. These surveys offer a gold mine for the study of micropolitics – the political choices and actions – of citizens in Germany and Britain and by extension, we will argue, in other established democracies.

As household surveys, they provide information on the most intimate of social units, but they offer few details on other forms of social life. There are no direct questions on other social networks or discussion groups. In Chapter 6, we show a strong association between living with others and turnout, net of the effects of a host of personal characteristics that keep people at home. BHPS asks a series of questions about the quality of relations among members of households, but we have found the responses difficult to interpret and to use as operational measures of critical concepts; all told, they are not helpful in the analysis of partisanship. GSOEP offers a few questions on policy preferences, and the British survey includes a

Preface

battery of such questions and also asks about political ideology. Because of potential problems of endogeneity with partisanship, we include few of them in our analyses (Anderson, Mendes, and Tverdova 2004; Erikson 2004; Evans and Anderson 2004; Johnston et al. 2005). These surveys are wonderful resources for the analysis of households; they enable us to explore the effects of these – but only these – intimate social relations on political preferences and behavior.

THE VOLUME'S PLAN

Put simply and directly, we argue that the intimate social contexts of people's lives influence their partisanship (support, choice, and constancy), net of other factors, like political interest and more distant and abstract social locations and identifications, like social class and religion. The strength of this relationship in a particular dyad (husband and wife in Germany and mother and child in Britain, to note two examples that we explore) depends on other variables that characterize each case. It follows that turnout and electoral choice reflect the electoral behavior of a person's spouse or partner, parents, and children and their own partisan constancy, which, as noted, also responds to the behavior of family and household members. We also demonstrate that negative results – not naming or voting for a political party – are particularly strong. The presence of two persons in a household who do not support Party A/B ensures that the other person refrains from naming that party. The presence of two who support A/B raises the probability of preferring A/B well above the mean, but it does not guarantee it. Throughout the volume, we interweave the principles of the social logic of partisanship with the analysis of bounded partisanship in Germany and Britain.

Our study begins by presenting the general theoretical framework. In the first chapter, we present the social logic of partisanship, reviewing the intellectual history of the approach as well as detailing general and specific hypotheses that we use to analyze German and British partisanship and voting behavior in Great Britain. Chapter 2, the first empirical chapter, details partisanship: partisan support, choice, constancy, and switching in our two cases. Because there have been so many studies of partisanship, we want first to demonstrate that the data taken from GSOEP and BHPS provide a picture that is sufficiently different from what we already know to sustain an analytical puzzle. Finding that most people are bounded partisans establishes this claim. Most people do not behave as if naming a party reflects a social identification; there is little evidence that persons choose their parties after a series of calculations that seek to foster and reflect their self-interest. Here, we lay out the properties of partisanship.

Preface

Then we apply models that apply the social logic of partisanship. In Chapter 3, we use Heckman Probit Selection models in order to analyze partisan support and choice and Zero-Inflated Negative Binomial models of partisan preference and constancy. These offer a preliminary argument on behalf of the utility of the social logic of politics. We develop this part of the analysis in two stages. The next three chapters delve deeply into German and British households. In Chapter 4, we explore political relationships among household partners. How similar are they? What accounts for variations in the level of political similarity? Does one partner dominate the other? This chapter shows that household cohesion with regard to politics is a variable, not a given, and that each partner influences the other. Following that, we detail the partisanship of young persons and examine the reciprocal influence of wives/mothers, husbands/fathers, and children on partisanship and electoral behavior. Chapter 5 looks at the components of partisanship alone and displays the central role of wives/mothers in the distribution of partisan choices within households. Chapter 6 extends the analysis to voting in British elections. In that chapter, we show that both partisan constancy and the electoral decisions of household members influence turnout and electoral choices. There is, we demonstrate, a social logic to partisanship and electoral behavior.

In the concluding chapter, we move our analysis beyond the two cases at hand to examine broader implications of the social logic of politics. Whereas we expect the general principles to apply to all political choices, political structures and institutions influence the particular patterns. Two dominant parties compete in both Germany and Britain, and their presence, campaign appeals, and activity necessarily influence the partisan choices and constancy of citizens. In the conclusion, we extrapolate our findings to people in similar party systems. Our ability to generalize to other established democracies is conditioned by variations in the party systems.

I

The Social Logic of Partisanship: A Theoretical Excursion

> [A]nd man became a living being, – with a soul able to speak and reason.
> Targum Onkeles on the phrase *l'nefesh hayah* in Genesis 2:7
>
> [M]an alone of all the animals is furnished with the faculty of language.
> Aristotle, *The Politics*, II: 10 (Barker 1962, 6)
>
> As a social being the person needs to be capable of reading messages from other persons, or responding to these and of composing intelligible messages to send out.... This person has to have beliefs about how the world is and how it works, ontological knowledge, and knowledge about how other persons behave.... Equipped with the wherewithal to make choices, the rational social being will apply choice to dealings with other persons and will develop strategies for manipulating and controlling them and for escaping unwanted control.
> Douglas and Ney (1998, 89)

Partisanship is a socially derived choice. In this statement, we draw together two divergent theoretical sources: a socialized understanding of humans and a perspective that emphasizes persons as beings that seek to make reasoned decisions. Our perspective denies that social groups subsume individuals; people do more than simply act as reflections of their social categories and locations. As important, persons are also not "free floating atoms" (to recall Marx's criticism of liberal theory). People reason, but they do not necessarily maximize expected utility or make the objectively correct choices. Instead, they employ subjective or bounded rationality, seeking to do the best they can with the intellectual and informational means available to them. Learning from others aids decisions, but learning from others is also what people do, as people.

Purposefully as well as inadvertently, people send and receive political messages. They influence those who receive their decisions and are, in turn, swayed by people whose messages they receive. The frequency of interaction affects the probability of influence: as more messages are sent and received, the level of influence shared increases. The level of shared trust affects political communication as well. Families, it follows, play a major role in political choices. These relationships are probabilities, not determined certainties. They apply to everyone. Politicians, whose political choices are made in contests with clear goals and rules, operate under an additional set of decision rules. We return to the distinction between citizens and politicians in the final chapter (Meehl 1977 and Riker 1982 provide the classic basis for this distinction). Here, we reason about citizens who are not politicians, and in the empirical analyses, we examine how they influence each other's choices about political parties.

THE SOCIAL LOGIC OF PARTISAN CHOICE

Inherently social, people live their lives by interacting with others, by considering and anticipating the behavior of others, and by influencing each other. Numerous social mechanisms account for this principle. Some emphasize the simple number of interactions: the greater the number, the more likely there is to be influence. Physical propinquity matters in Tobler's Law, "Everything is related to everything else, but near things are more related than far things" (cited in Miller 2004). In turn, the ancient principle of "like to like" (i.e., similar persons are drawn to each other) and its obverse (i.e., "opposites attract") also draw attention to the nature of the interactions.

Consider also the following general statements taken from diverse research traditions:

A social being has one prime need – to communicate. Because it is a social being, everything in its genetic inheritance, especially its intelligence, must be equipped to read the signals and to signal back to beings of its kind. (Douglas and Ney 1998, 46)

Among social species we are unique in our plasticity.... Behavioral rules (including social norms and conventions) make social interaction predictable, so that interdependent individuals can influence one another in response to the influence they receive, hereby carving out locally stable patterns of interaction. In short, rules are not simply analytic shortcuts that lower the cognitive costs of decision-making. They are the grammar that structures our social life. (Macy 1998, 221)

The capacity of people to engage in complex cooperation, and to seek and give social support in particular, might be a defining characteristic of our species. (Cunningham and Barbee 2000, 274)

Individuals learn appropriate social behavior from observing each other. This proposition emphasizes the interconnected nature of social life, underlining the centrality of observing and copying others as people perceive, evaluate, and make decisions about how to act. We view behavior as more correct in a given situation to the degree that we see others performing it. (Cialdini, cited in Axelrod 1997b, 58)

Social Logic of Partisanship

> Conformity is inherited. It is inherited *socially*, because social norms develop that reinforce conformity –glory [sic!] prestige, pride in the group, patriotism.... It is inherited *genetically* because social learning provides such an evolutionary advantage to the individual. (Jones 2001, 117)

Consider as well the recent research that has located in the dorsal striatum section of the brain the physical location for these kinds of expectations and learning rules (Berns et al. 2005 and King-Cassas et al. 2005). Diverse theories agree that humans are social beings necessarily living social lives.

Social animals though they be, humans do not move in herds (though one should not minimize the importance of cue-giving and taking among animals). Even as they communicate, persons stand apart from even the closest associates, spouses, and parents; each and all always retain the ability to accept or reject cues, to adhere to or ignore other people's expectations and preferences. More, members of social groups disagree, sometimes sending conflicting messages to each other, sometimes fighting. Also, persons belong to many social groups, whose signals vary in the extent to which they are consonant with each other. Even as a person is influenced by members of a group, he or she always chooses whether or not to follow others, and the more that the members of a person's groups disagree with each other, the more that he or she must pick whom to follow. The fundamental need to communicate does not remove the ability to stand apart from others and to ignore or disagree with them, and, therefore, the need to decide to agree or to disagree. And so, political influence, like other social phenomena, is best understood as a probabilistic relationship.

Seeking to account for the extent and ways that individuals learn from each other, an assortment of research traditions underlines the variability of social conformity. Some make this argument on theoretical grounds, and here too the overlap between divergent theories underscores the power of the point (for Freudian theories see Alford 1994; Bion 1961; for rational choice theories, see Lichbach 1996; 1997; Olson 1965). Some, like Mary Douglas, join theory and a synthesis of anthropological studies of small-scale communities (Douglas 1986; Douglas and Ney 1998). Political scientists display consistent evidence of political diversity in discussion networks and families (Huckfeldt, Johnson, and Sprague 2004; 2005; Stoker and Jennings 2005; Zuckerman and Kotler-Berkowitz 1998; Zuckerman, Fitzgerald, and Dasović 2005). Even as some argue for the existence of a powerful drive to cultural agreement (Axelrod 1997a; 1997b), classic theories (Simmel 1955), and recent work using agent-based models (Johnson and Huckfeldt 2005) sustain the principle of opinion diversity within members of social networks. Social conformity may not be assumed; individuals are not subsumed by their social ties; surrounded and influenced by others, they make their own decisions; relationships of social influence are probabilities.

How do they decide? The principles of bounded or subjective rationality guide the way. People seek to make what they perceive to be the correct decision, even if they do not succeed and even if the decision is not objectively correct (see Boudon 1992; 1998; Gigerenzer and Selten 2001a; Jones 2001; and for the classic sources on bounded rationality see Simon 1965 [1957]; 1999). They seek strong reasons for what they do (Boudon 1992; 1998).

Usually eschewing complex calculations, they apply useful rules of thumb – or usually correct theories or understandings or decision heuristics, which are conditioned by their social circumstances and their intellectual abilities. Many of these decision rules involve taking cues from members of one's immediate social environment. Many do not. Here, we emphasize those that draw on principles of social learning for decision heuristics.

We begin with the teachings of Kenneth Arrow. In large publics, he argues, the rationality assumption does not apply. Individual rationality is not just a property of each person. It is linked to the macro characteristics of the economy, the presence of economic equilibrium, perfect competition, and "completeness of markets:"

> When these assumptions fail, the very concept of rationality becomes threatened, because perceptions of others and, in particular, their rationality become part of one's own. (1986, S389)

Consider also the observations of other economists and decision theorists.

> [I]n social species, imitation and social learning can be seen as mechanisms that enable fast learning and obviate the need for individual calculations of expected utilities.
> Social norms can be seen as fast and frugal behavioral mechanisms that dispense with individual cost-benefit computations and decision-making. (Gigerenzer and Selten 2001a, 9–10)

> Social imitation can help make decisions with limited time and knowledge. Heuristics such as "eat what your peers eat" and "prefer mates picked by others" can speed up decision making by reducing the time spent on information search. (Selten 2001, 48)

> Imitate if Better, assumes that individuals imitate all others who are more successful than themselves, and stick with their current strategy otherwise.
> Proportional Imitation – which dictates imitating those who are more successful than oneself with a probability that is proportional to the difference between the observed and current degrees of success. (Group Report in Gigerenzer and Selten 2001a, 175)

Social Logic of Partisanship

The behavioral decisions made by animals are to a large part influenced by what other animals are doing. Social learning is not restricted to humans, or clever animals, but is a fundamental feature of vertebrate life. In an array of different contexts, numerous animals adopt a "do-what-others-do strategy," and in the process, learn an appropriate behavior. (Laland 2001, 233)

Most people most of the time eschew optimization and frequently apply social learning as a strategy for making choices.

How do boundedly rational persons engage in social learning? They are influenced by persons from whom they take other cues; successful learning in the past induces the expectation of successful learning in the present and future. When cue-givers disagree with each other, boundedly rational persons choose alternative strategies. They may follow the net preferences of these persons; they may weight them on an easily constructed scale; they may follow the most knowledgeable, the most trusted, the most frequently seen, or they may apply some other decision heuristic. No single strategy always applies. In the empirical portions of this book, we will show that this variation even characterizes families and households.

Joining these two themes implies two explanatory mechanisms. 1. People learn from others because that is what social beings do. 2. People learn from others because it is an effective and frequently used decision heuristic. Both general propositions justify the claim that partisanship and electoral choice are socially derived choices, such that the preferences of others affect the decision to support and vote for a political party. They imply too that family members have a particularly strong influence on each other's political decisions. Note as well the limits of these claims. They imply no more than people sometimes learn from others and sometimes do not. The relationship is probabilistic. Like other social mechanisms, these are sometimes true theories (Elster 1998; Hernes 1998).

From whom are persons especially likely to take political cues? Individuals are especially likely to follow those persons from whom they take other cues, who know more than they do, those on whom they depend, whom they trust, with whom they regularly interact, and whom they perceive as being like themselves. They are especially likely to follow those to whom they feel accountable (Tetlock and Lerner 1999). Conversely, they are especially likely to reject the political cues of persons who know less than they do, whom they usually ignore, perceive as different from themselves, whom they do not trust, and who are strangers to them, and to whom they do not feel accountable. More concretely, in the following chapters, we show that family members are more likely to influence partisan choice than, for example, membership in a trade union or religious congregation. Note that we do not emphasize the importance of asymmetries in political knowledge or expertise as sources of political influence. In our view, choosing a political party is not an optimizing decision for which additional knowledge might affect the outcome. Trust and frequency of interaction underpin the political influence that affects this choice.

Consider, however, that the influence of social intimates on partisanship does not determine that they incline towards the same party; each or all may opt not to prefer any party at all. Indeed, in the extreme case, when persons who live together spend all their time together, they are social and political isolates and, therefore, are likely to support no party. Interacting only with each other, they only know each other; they take interest in no one else. Persons with ties outside the household have opportunities to absorb political messages from those with whom they work, play, pray, and share membership in social organizations. In turn, the cues that move across these weak ties affect partisan preferences and then are transmitted to those with whom they share strong ties.

POLITICAL LEARNING IN POLITICAL SCIENCE

Our understanding of the dynamics of political choice overlaps with and departs from other studies that also accept the principles of social learning. Consider first how we stand in relation to Achen's (2002) application of rational choice theory to partisan socialization within families. Like Achen, we accept the principle that persons are not passive receptacles of political messages. Like Achen, we maintain that persons learn from those around them. We differ, however, with regard to the process of political learning. For Achen, children learn from their parents by following a complex set of steps. Children, like all persons, support a political party, "when they believe that its future course of benefits exceeds that of the other alternatives. When the voters expect that a party will benefit them in the future, they will be said to 'identify' with that party" (Ibid, 153). Each party is perceived to offer "a stream of benefits (cardinal utilities) that varies randomly around a constant mean" (Ibid, 153–4). Furthermore, children, like their parents and everyone else, take their understanding of desired political benefits from their social positions (Ibid, 154). And so Achen asks and answers the fundamental question of political socialization: "Why should children decide to think like their parents? The answer given is that parent and child will often occupy similar positions in the social structure, and thus parental experience is likely to be relevant to the child's future life" (Ibid, 154–5). Seeking to maximize their expected benefits and supposing that they will occupy the same social positions as their parents, young people take their parents' partisanship.

Social Logic of Partisanship

As noted, we eschew the claim that people choose parties by calculating costs and benefits: choosing a party carries no personal outcome; benefiting no one, partisanship suggests no interests, and the person who can make the required complex calculations of optimizing choices can easily recognize that these are not expectations about outcomes but hopes and wishes. More generally, as citizens consider the political parties they lack a "macroscopic driving force" (Aumann 2005) that guides their decisions. Choosing a political party is nothing like making money in the market place or winning in chess; for that matter voting too bears none of the critical characteristics that imply the strategic importance of optimization. Because partisanship and electoral behavior lack clear goals, the classic principles of rational choice theory do not apply. Furthermore, our analysis also shows that social class is a much weaker predictor of partisan choice and consistency than Achen's assumptions allow.

Children learn partisanship from their parents, we suggest, not because they expect to benefit from the choice and not because they and their parents share the same social location, and so it is smart to learn from their parents. Rather, interacting frequently with their parents, they take many cues, among them political ones, from them. Parents are generally trusted (indeed usually beloved) persons, with whom young persons interact on a regular basis. And so, when young people offer a positive answer to a question about party preference, they usually echo what they have heard at home. In Chapter 5, we show that they do so in specific ways: almost always ignoring or rejecting the party that their parents do not support and varying the rate by which they accept their parents' partisan preferences. Furthermore, as children enter and move through their third decade, obtaining the right to vote, social relationships within households imply that parents also learn from their children. In Chapters 5 and 6, we show that the exchange of partisan influence is strongest between mothers and children.

Indeed, one of the questions that helps to define Achen's approach further elucidates our differences, and, we suggest, highlights the strength of our approach. After denying that children are passive recipients of their parents' wisdom, he continues, "Parents are rarely able to influence their teenage children's hairdos, clothing styles, tastes in popular music, or even more important decisions like the choice of a life partner. Why should party identification be any different? Put more precisely, why do teenagers implicitly accept their parents' advice about political parties while they avoid taking it on a great many other topics?" (Achen 2002, 152). Achen's model answers these questions. We are less surprised than Achen that young persons are more likely to accept their parents' partisanship than that of their friends. After all, parents are more likely to have partisan preferences than are the youngster's peers. Parents are also more likely to display their political preferences in words and gestures than are young persons, among whom political discussions are not frequent; teenagers rarely exchange political cues. And why suggest, as does Achen, that partisanship is unique? In Chapter 5, we show that children usually take the religion and the social class preference of their parents, but are also more likely to deny an affiliation or identification with any of the relevant social groups. There is nothing unique about partisanship, and partisan preference is not like the choice of clothes or hairdos; it is more like the choice of religion. In Chapter 5, we summarize a substantial literature in social psychology that maintains that parents usually matter more than friends, and we elaborate our argument on political learning in households.

Consider also how we stand in relation to others who apply decision heuristics to political choice. Like Fowler (2005), we expect each person's decision to vote to influence and to be affected by the turnout of others in their social networks. Also like Fowler, we emphasize exchanges in families and households, rather than learning from those who are very knowledgeable about or interested in politics (Downs 1957; Huckfeldt, Johnson, and Sprague 2004), the media (Mutz 1998), or political institutions (Lupia and McCubbins 2000). We agree with Lupia and McCubbins that social learning requires trust, but we believe that it is both more parsimonious and more accurate to look to loved ones – parents, spouses, children, friends, and workmates – than to political institutions for sources of trusted information. The political preferences of one's social intimates provide the lens through which a person perceives the political parties and the general political world (Schmitt-Beck 2003). All these analyses differ from those that apply decision rules on partisanship that are internal to the actor, such as ideology (Downs 1957; Boudon 1992), perceptions of the incumbent (Boudon 1992; Sniderman, Brody, and Tetlock 1991, 180), and assessments of relative feelings about groups – likeability or evaluations of desert, the desert heuristic (Boudon 1992, 23; Sniderman, Brody, and Tetlock 1991, 87–8). We emphasize personal ties, especially frequency of interaction, past learning, and trust as the factors that enable persons to apply heuristics to political choice.

In this volume, we present the results of several analyses that demonstrate the importance of social factors as predictors of partisanship. Both immediate social circles (household partners and parents and children) and more distant social contexts (social class and religious membership and identifications) as well as the political interest of the respondent and others in the social network influence whether or not a party is chosen, the party named, and the consistency of that selection. This evidence illustrates the power of the social logic of partisanship.

AN INTELLECTUAL HISTORY OF THE SOCIAL LOGIC OF POLITICS

Our analysis of partisanship[1] rests on what we call the social logic of politics. In this section, we recount an intellectual history in order to provide theoretical underpinning for a set of related propositions: (1) The social logic of politics has an ancient pedigree that supports modern social science and is part of most people's intuitive understanding of social life. The principles sustained the first generation of studies of partisanship. (2) The reasons to move away from these principles were weak when first offered and are no longer tenable at all. (3) There are strong theoretical reasons to return to the social logic of politics, in general, and as applied to the analysis of partisanship in particular.

Students of politics have always known that the immediate social circumstances of people's lives affect their political perceptions, preferences, and behavior. Recurrently, however, issues of theory and method induce them to follow other paths, and, recurrently too, they come back to this perspective.

Consider first the intellectual pedigree. Relevant stories, principles, themes, and statements abound in the Bible, Greek classics, and foundational works of the medieval period. The Bible opens (and see the reference to Onkeles's commentary at the start of the Preface) with stories that center the principles of human life on the family. Aristotle begins the *Politics* with a discussion of the household, citing Hesiod: "First, house, and wife, and ox to draw the plow" (Barker 1962, 4). In more abstract terms, Aristotle defines the family as the nucleus of the polity, a claim that resonates through centuries of Christian political thought. Even a cursory look at these literatures finds the family at the heart of personal and collective life.

More directly linked to the task at hand, these literatures recognize that people live their lives by learning from those around them. Ancient proverbs are presented as analytical truths and behavioral rules that lead to a good life. One set of principles involves the benefits obtained from good friends and the need to avoid bad associates. Another set involves the selection of colleagues: similar persons associate with each other; but sometimes they repel each other, and different people may find themselves drawn together as well.

---

[1] Portions of this section are taken from Zuckerman (2005b).

Partisan Families

Here we present Aristotle's views, highlighting critical statements and propositions.[2] Aristotle explores the importance of friends in the following passage from the *Nicomechean Ethics*:

> No one would choose to live without friends, even if he had all other goods.... Friends help young men avoid error; to older people they give the care and help needed to supplement the failing powers of actions which infirmity brings in its train; and to those in their prime they give the opportunity to perform noble actions.... Friends enhance our ability to think and to act.... There are, however, several controversial points about friendship. Some people define it as a kind of likeness, and say that friends are those like us; hence according to them, the proverb: "Like to like," "Birds of a feather flock together." Others, on the contrary, hold that all similar individuals are mutually opposed. (Aristotle, *Nicomechean Ethics* 1155a, 214–6)

> The friendship of base people becomes wicked, because unsteady as they are, they share in base pursuits, and by becoming like one another they become wicked. But the friendship of good men is good, and it increases with [the frequency of] meetings. Also, it seems, they become better as they are active together and correct one another: from the mould of the other each takes the imprints of the traits he likes, whence the saying: "Noble things from noble people." (Ibid, 1172a, 271–2)

These proverbs did not originate with Aristotle. Rather, he cites Homer, Empedocles, Euripides, and Heraclites. Indeed, these proverbs indicate that even Aristotle considered the aphorisms "like to like," "birds of a feather flock together," and "opposites attract" to be the wisdom of the ancients.

Centuries later, Maimonides (1135–1204) and Aquinas (1225 or 1227–74) extend these principles:

> It is natural to man to be influenced by the beliefs and practices of his friends and neighbors and to behave like the practices of his community. Therefore, a man should attach himself to the righteous and sit always among the wise, so that he may learn from their deeds. He should also stay away from evil people who walk in darkness, so as not to learn from them. (Maimonides, *Mishneh Torah*, 6:1, our translation)[3]

---

2 The Hebrew Bible and the *Iliad* too offer similar principles for the good life. For example: "Happy is the man, who has not followed the counsel of the wicked, or the path of sinners, or joined the company of the insolent; rather his concern is the teaching of the Lord, and he recites that teaching day and night" (Psalms 1:1). "He who walks with the wise will become wise, and he who gathers with the fools will become bad" (Proverbs 13:20). "God always pairs off like with like" (*The Iliad*, XVII, 1:218).

3 In *The Guide of the Perplexed*, Maimonides defines humans as beings that live, reason, and die, and equates reason and the ability to speak (1963, 455).

Social Logic of Partisanship

Friends become better by working together and loving each other. For one receives from the other an example of virtuous work which is at the same time pleasing to him. Hence it is proverbial that man adopts noble deeds from noble men. (Aquinas, *Commentary on the Nicomechean Ethics* 2:5)

Are friends necessary for happiness? *Response*: If the question refers to the happiness of the present life then, as the Philosopher says, the happy man needs friends, not because they are useful, since he is able to get along without help, nor for the enjoyment, since the activity of virtue furnishes him with complete joy, but for good activity, that is so that he may do good to them and delight in seeing them do good and be helped by them in doing his own good deeds. For, in order to do well, whether in the works of the active life or in the activity of the contemplative life, man needs the help of friends. (Aquinas, *Treatise on Happiness*, 12)

The Bible, Greek wisdom, and the sources that develop and bring them together highlight a basic understanding of social life. Persons need and rely on each other; they learn from each other, and they enable each other to live well. As important, people usually associate with others like themselves, but they sometimes avoid their similars, choosing instead to bond with their opposites. The ancients understand the selection of associates as inherently probabilistic.

Among the principles offered by the founders of modern social science too, it is easy to find statements that affirm the propositions of the social logic of politics. Tocqueville conceptualizes individualism by melding person and primary group (1969[1958], 508), and, he argues, in democracies people take their political views from those around them (Ibid, 643). Marx maintains that as individuals interact with others at the same relationship to the means of production and, therefore, the same social circumstances, they will come to share the same perceptions, understandings, values, and political action. As these processes unfold, these persons form a social class – a political association in opposition to the other social class, which takes on ontological reality (for the appropriate passages in Marx, consult Bottomore and Rubel 1956, and for useful analyses see Katznelson 1986; Zuckerman 1991). Max Weber replaces the deterministic nature of Marx's analysis with probabilistic relationships (see for example Weber 1978 [1922] II, 928–30). As important, Weber also places the inherent sociality of people's lives at the heart of his social science (Weber 1978 [1922] I, 4, 26, and see the opening to the book's Preface). For Emile Durkheim, the interactions of persons in groups define their essential humanity (Durkheim 1966 [1933], 26), and Gaetano Mosca refers "to mimetism...the great psychological force whereby every individual is wont to adopt the ideas, the beliefs, the sentiments that are most current in the environment in which he has grown up" (Mosca 1939, 26). Georg Simmel uses networks of social interaction to explore the tension between persons and groups (Simmel 1955, 140, 141, 151). Differences appear among

Partisan Families

the titans of social science. Marx and Durkheim depict groups as part of the social ontology. Tocqueville, Weber, Mosca, and Simmel examine interactions among persons, maintaining analytical space for individuals and groups. The founders of social science offer theoretical and conceptual reasoning and some empirical evidence for the social logic of politics.

THE SOCIAL LOGIC OF POLITICS IN THE FIRST-GENERATION OF STUDIES OF PARTISANSHIP

Returning to the foundational texts of the behavioral revolution in political science highlights the fundamental importance of the social logic of politics for the first generation of studies of partisanship. We begin with a selection from *The American Voter* (Campbell, Converse, Miller, and Stokes 1960). This volume raises the flag of the Michigan School of electoral analysis, institutionalizing the research agenda for electoral studies in the United States and other established democracies. It spawns as well the American National Election Surveys and parallel studies elsewhere. This research has provided the lion's share of evidence for the study of partisanship, during the past fifty years.

Campbell et al. recognize the impact of immediate social circles on the ways that persons perceive and act in politics:

Not only does the individual absorb from his primary groups the attitudes that guide his behavior; he often behaves politically as a self-conscious member of these groups, and his perception of their preferences can be of great importance for his own voting act. Our interviews suggest that the dynamics of these face-to-face associations are capable of generating forces that may negate the force of the individual's own evaluations of the elements of politics. Probably this happens most often in the relations of husband and wife. (Ibid, 76)

Knowledge of social processes may add much to our understanding of the fact that party allegiances not only remain stable but grow stronger over time. In addition to intra-psychic mechanisms that act in this direction, social communication in a congenial primary group may constitute a potent extra-psychic process leading to the same end. The ambiguity of the merits of political objects and events is such that people are dependent on "social reality" to support and justify their political opinions. When primary groups engage in political discussions and are homogeneous in basic member viewpoints, the attitudes of the individual must be continually reinforced as he sees similar opinions echoed in the social group. (Ibid, 293)

Consider too the views of their colleagues, beginning with Robert Lane:

Political participation for an individual increases with (a) the political consciousness and participation of his associates, (b) the frequency and harmony of his interpersonal contacts and group membership, and (c) the salience and unambiguity of his group references. (Lane 1959, 189)

Social Logic of Partisanship

Groups orient a person in a political direction specifically by (a) redefining what is public and private in their lives, (b) providing new grounds for partisanship. (Ibid, 195)

Even V. O. Key, who was among the first to argue against the social logic of politics, accepts its general importance and particular significance with regard to the political parties. "Probably it is correct to picture the political system as one in which a complex network of [primary] group relations fixes and maintains opinions in some systematic relation to the larger components of the system, such as political parties" (Key 1961, 69–70). Indeed, we return to the reciprocal relationship between political exchanges in primary groups and the political parties in the final chapter. Sidney Verba underlines the importance of small groups in the analysis of political processes as well as the behavior of individuals:

If we are to understand the political process, greater consideration must be given to the role of face-to-face contacts. Primary groups of all sorts mediate political relationships at strategic points in the political process. They are the locus of most political decision-making, they are important transmission points in political communications, and they exercise a major influence on the political beliefs and attitudes of their members. (Verba 1961, 4)

It is well known that the face-to-face groups to which an individual belongs exert a powerful influence over him; that he will accept the norms and standards of the group.... [This] is one of the best documented generalizations in the small group literature. (Ibid, 22–3)

And finally, note as well Heinz Eulau's general principles:

Just as the significant environment of the individual is another individual, so the significant environment of the group is another group. (Eulau 1962, 91–2)

Political behavior is likely to vary with the type of groups in which the individual is involved. (Eulau 1986, 38)

The behavioral revolution in political science begins with the principles of the social logic of politics.

Every revolution draws on, negates, and transforms what precedes it, and so too in political science. Of direct and powerful relevance to these political scientists is a group of electoral sociologists at Columbia University, led by Paul Lazarsfeld and Bernard Berelson. Robert Merton and Edward Shils offer more general theoretical statements. At the same time, Campbell, Converse, Miller, and Stokes reinterpret Kurt Lewin's social psychology and Robert Merton's reference group theory in order to reformulate the understanding of the political relationship between the group and the individual. The political scientists accept as well Leon Festinger's analysis of the relationship between the individual and the political reality portrayed by his or her peers. All of these sources accept, apply, and develop the principles of the social logic of politics. And so the behavioral revolution in the study of political behavior transforms these intellectual sources.

SOURCES IN SOCIOLOGY

Paul Lazarsfeld and his colleagues at Columbia University, Bernard Berelson, Hazel Gaudet, and William McPhee, are the first to link the analysis of the social logic of politics to the study of electoral choice. *The People's Choice: How the Voter Makes Up His Mind in a Presidential Campaign* (Lazarsfeld, Berelson, and Gaudet 1968 [1948] based on the 1940 election) and *Voting: A Study of Opinion Formation in a Presidential Campaign* (Berelson, Lazarsfeld, and McPhee 1954) also introduce mass surveys into the analysis of political preferences. These studies interview respondents several times during the life of an electoral campaign, offering the first panel surveys of electoral behavior. Several sets of questions ask for information on the members of their immediate social circles, family, friends, workmates, and neighbors. These social scientists examine evidence drawn from single communities, Elmira, New York, and Erie County, Pennsylvania, not a nationally representative sample of the electorate. The electoral sociologists initiate a research path, which Campbell, Converse, Miller, Stokes, Key, Eulau, and other political scientists follow and then redirect.

How important is the research of the electoral sociologists for the behavioral revolution in political science? Here are the opening words of *The American Voter:*

In the contemporary world the activity of voting is rivaled only by the market as a means of reaching collective decisions from individual choices.... Indeed, anyone who reads the literature of voting research must be impressed by its proliferation in recent years. The report of one major study lists 209 hypotheses about voting in political elections, which recent work has tended to confirm. (Campbell et al. 1960, 3)

Here, the volume's first footnote cites Berelson, Lazarsfeld, and McPhee's *Voting*. Similarly, the first sentence of Key and Munger's classic article cites *The People's Choice*, as they frame their presentation in opposition to Lazarsfeld and his colleagues: "The style set in the Erie County study of voting, *The People's Choice*, threatens to take the politics out of the study of electoral behavior" (Key and Munger 1959, 281). Heinz Eulau's first book also begins by echoing Key and Munger's point, citing the same passage from *The People's Choice* and declaring his opposition to social determinism (Eulau 1962, 1). The leaders of the behavioral revolution in political science first examine electoral choice through the lenses of electoral sociology.

Social Logic of Partisanship

As a result, they begin with the principles of the social logic of political behavior. How do Lazarsfeld, Berelson, Gaudet, and McPhee enunciate those explanatory mechanisms? What are the leaders of the behavioral revolution in political science reading? And to what precisely do they react? Consider some critical themes in *The People's Choice*:

> While the individual preserves his security by sealing himself off from the propaganda which threatens his attitudes, he finds these attitudes reinforced in his contacts with other members of the group. Because of their common group membership, they will share similar attitudes and will exhibit similar selective tendencies. But this does not mean that all of the members of the group will expose themselves to exactly the same bits of propaganda or that they will be influenced by precisely the same aspects of common experiences. (Lazarsfeld et al. 1968, xxxii)

The boldest version of this statement sounds like social determinism, and, as we show later, it provides a point of attack for the political scientists:

> There is a familiar adage in American folklore to the effect that a person is only what he thinks he is, an adage which reflects the typically American notion of unlimited opportunity, the tendency toward self-betterment, etc. Now we find that the reverse of the adage is true: a person thinks, politically, as he is, socially. Social characteristics determine political preference. (Ibid, 27)

It is clear, however, that the last line only highlights the general stance; it is not a theoretical principle. The volume abounds in statements and evidence, which maintain that the effects of social context on political preferences vary (see for example Ibid, 137 and 150). It demonstrates how personal contacts affect the electoral choices of undecided citizens. Several factors drive the process: the power of the two-step flow of communications; personal contacts that need no particular purpose; flexibility when encountering resistance; rewards of compliance; greater level of trust in the source; and persuasion without conviction (Ibid, 150–7):

> In short, personal influence, with all its overtones of personal affection and loyalty, can bring to the polls votes that would otherwise not be cast or would be cast for the opposing party just as readily if some other friend had insisted. (Ibid, 157)

In a footnote in the book's last chapter, Lazarsfeld and his colleagues sketch a research project that would examine variations in the political homogeneity of social groups:

> The statement that people vote in groups is not very satisfactory. People belong to a variety of groups and therefore further research is necessary on the question: with *which group* are they most likely to vote? (Ibid, 170)

The complexity of social ties affects the political cohesion of social groups.

Partisan Families

Berelson, Lazarsfeld, and McPhee develop this perspective. They locate high levels of political homogeneity in primary groups (1954, 88–118). During an election campaign, they find voting intention responds directly to a combination of cues and requests from members of discussion circles, families, friends, and co-workers (Ibid, 118–49). Finally, they add that the logic of democracy works at the aggregate level, not the individual level:

> True, the individual casts his own personal ballot. But as we have tried to indicate throughout this volume, that is perhaps the most individualized action he takes in an election. His vote is formed in the midst of his fellows in a sort of group decision – if indeed, it may be called a decision at all – and the total information and knowledge possessed in the group's present and past generations can be made available for the group's choice. Here is where opinion-leading relationships, for example, play an active role.
>
> Second, and probably more important, the individual voter may not have a great deal of detailed information, but he usually has picked up the crucial *general* information as part of his social learning. (Ibid, 320–1)

Working with Elihu Katz, Lazarsfeld presents a broader analysis of the process by which people form attitudes, preferences, and values (Katz and Lazarsfeld 1955). Lazarsfeld and his colleagues articulate the principles of the social context of politics as testable hypotheses. They examine panel surveys of particular communities in order to test and demonstrate the power of this theoretical perspective.

These sociologists draw directly on Simmel's and Weber's social science. They complement the sociological wisdom of their own day as well. Robert Merton highlights the centrality of primary groups as he develops the concepts of reference group theory. He maintains that the Elmira election studies confirm the theoretical, normative, and empirical claims of pluralist theory:

> [I]t is not "individuals," tacitly conceived as "sand heap [sic!] of disconnected particles of humanity," who are protected in their liberties by the associations which stand between them and the sovereign state, but "persons," diversely engaged in primary groups, such as the family, companionships, and local groups. That figment of the truly isolated individual, which was so powerfully conceived in...Hobbes' *Leviathan*, and which was since caught up in the assumptions of the liberal pluralists, is a fiction which present-day sociology has shown, beyond all reasonable doubt, to be both untrue and superfluous....
>
> [E]ven the primary groups in which persons are in some measure involved do not have uniform effects upon the orientations of their members.... Moreover, when conflicting value-orientations obtain in the primary-groups, and the modal orientations of the larger social environment are pronounced, the mediating role of the primary group becomes lessened or even negligible, and the influence of the larger society becomes more binding. (Merton 1957, 334–5)

Social Logic of Partisanship

Again, we find the general principles: people depend on each other, and there is a complex relationship among individuals, primary groups, and the broader society. This perspective extends beyond Merton, Lazarsfeld, and their colleagues at Columbia University. It appears in Edward Shils's classic essay on primary groups (1951, 69) and David Riesman's *The Lonely Crowd* (1961 [1951], 23). The founders of contemporary sociology refurbish and transmit the principles that began with the ancients and were presented again by Tocqueville, Marx, Weber, Durkheim, Mosca, and Simmel. The electoral sociologists and their colleagues offered political scientists a choice: follow us or blaze your own path.

SOURCES IN SOCIAL PSYCHOLOGY AND DECISION THEORY

Social psychologists – especially Kurt Lewin and Leon Festinger – provide another source for ideas that guide the behavioral revolution in political science. The authors of *The American Voter* model their "funnel of causality" on Lewin's field theory. This analytical approach applies a large set of immediately relevant explanatory factors (Campbell et al. 1960, 33). Key (1961, 62) and Verba (1961, 23) account for the tendency toward conformity in political preferences among members of primary groups by referring to Festinger's work on cognitive dissonance, and Campbell and his colleagues draw on Festinger to raise questions about the reliability of the respondents' reports to describe their immediate social circles. As the political scientists analyze political preferences, they again utilize the principles of the social logic of politics.

Lewin's social psychology argues for the utility of examining groups as collectives, defined by the interdependence of members:

> Conceiving of a group as a dynamic whole should include a definition of group which is based on interdependence of the members.... A group, on the other hand, need not consist of members which show great similarity.... Not similarity but a certain interdependence of members constitutes a group....
>
> [E]ven a definition by equality of goal or equality of an enemy is still a definition by similarity. The same holds for the definition of a group by the feeling of loyalty or of belongingness of their members. (Cartwright 1964 [1951], 146–7)

Abstract categories like social class, ethnicity, or religion, therefore, do not define social groups. By implication, sharing identification with a political party does not define a group. Indeed, psychological attachment provides the weakest form on which to base a group. It applies because it might constitute "a certain kind of interdependence, because there might be interdependence established by the feeling" (Ibid, 146–54, and see also Lewin 1948, 84ff.). Analyzing an individual, Lewin maintains, requires examining the person's "life-space," which is defined as anything that might affect the person. One segment includes the individual's perceptions; another examines members of a person's immediate social circle (see Lewin 1964 [1951], xii). Here too, the immediate social circumstances of people's lives affect their perceptions, choices, and actions.

Leon Festinger offers a tripartite analysis of the individual and the social group. Opinions, preferences, and beliefs are a joint function of how "real" the matter is, the views held by the members of a person's group(s), and the person's own conception(s):

Validity of opinion depends on what others around him say: "An opinion, a belief, an attitude is 'correct,' 'valid,' and 'proper' to the extent that it is anchored in a group of people with similar beliefs, opinions, and attitudes." (Schachter and Gazzaniga 1989, 119)

In turn, Festinger recognizes a fundamental tension in the set of explanatory mechanisms. Persons are influenced by members of their groups, and they join groups whose views conform to their own:

It is to some extent inherently circular since an appropriate reference group tends to be a group which does share a person's opinions and attitudes, and people do locomote *into* such groups and *out* of such groups which do not agree with them. (Ibid, 19)

Drawing on the results of his own and other studies, Festinger notes a strong tendency for members of groups to adopt similar views: "Belonging to the same group tends to produce changes in opinions and attitudes in the direction of establishing uniformity within the group." Furthermore, the amount of change toward uniformity is a function of how attractive belonging to the group is to its members (Ibid, 161).

Festinger's most important contribution, the theory of cognitive dissonance, combines these ideas into three core principles: (1) There may exist dissonant or "nonfitting" relations among cognitive elements. (2) The existence of dissonance gives rise to pressures to reduce the dissonance and to avoid increases in dissonance. (3) Manifestations of the operation of these pressures include behavior changes, changes of cognition, and circumspect exposure to new information and opinions (Ibid, 225). Any of these outcomes may appear at any time. Like the electoral sociologists, Festinger and Lewin articulate theoretical principles in line with the social logic of politics, and not surprisingly both also draw on Simmel's and Weber's social science.

Parallel developments in the study of political organizations echo these theoretical principles. Writing at about the same time, Herbert Simon also emphasizes the centrality of social context for understanding people's decisions. Consider a passage from the introduction to his classic; *Administrative Behavior*:

> Organization is important, first, because in our society, where men spend most of their waking adult lives in organizations, this environment provides much of the force that molds and develops personal qualities and habits. (Simon 1965 [1957], xv)

> In the pages of this book, the term *organization* refers to the complex pattern of communications and other relations in a group of human beings. This pattern provides to each member of the group much of the information, assumptions, goals, and attitudes that enter into his decisions, and provides him also with a set of stable and comprehensible expectations as to what the other members of the group are doing and how they will react to what he says and does. The sociologist calls this pattern a "role system:" to most of us it is more familiarly known as "organization." (Ibid, xvi)

In Simon's presentation, social context especially affects the initial decision, preference, or social action:

> Two principal sets of mechanisms may be distinguished: (1) those that cause behavior to persist in a particular direction once it has been turned in that direction, and (2) those that initiate behavior in a particular direction. The former are for the most part – though by no means entirely – internal. Their situs is in the human mind.... Behavior-initiating mechanisms, on the other hand, are largely external to the individual, although they usually imply his sensitivity to particular stimuli. Being external, they can be interpersonal – they can be invoked by someone other than the person they are intended to influence. (Ibid, 95)

In *Administrative Behavior* and *The American Voter*, immediate social circumstances are the source of a person's initial political preferences. In both works, social circles provide stability for political preferences. Simon's approach to the analysis of decisions emphasizes the context in which choices are made and the cognitive factors that limit people's ability to behave in ways consistent with rational choice theory.

TURNING AWAY FROM THE SOCIAL LOGIC OF POLITICS

Notwithstanding their initial theoretical stance, the founders of the behavioral analysis of political preferences and electoral choices institutionalize a research agenda that generally ignores the social logic of politics. They conduct surveys that examine individuals but ignore the members of their social circles, and they transform social groups into objects of individual identification. In the analysis of electoral decisions, they focus on political attitudes, perceptions of the candidates, and policy preferences. In *The American Voter*, social contexts provide background factors. As the

Partisan Families

American National Election Surveys becomes the primary source of data on electoral behavior for political scientists, they frame research around party identification and the issues and perceptions of the candidates of particular elections.[4] As this model travels across oceans, it structures electoral research in other democracies.[5] As Key and Anthony Downs introduce rational choice theory into the study of partisanship, they leave behind the principles of the social logic of politics. As debates rage between the Michigan School and proponents of rational choice theory, scholarly attention focuses on individuals. The social logic of politics loses scholarly prominence.

Several factors help to explain this change in direction. One set derives from the decision to use national sample surveys as the exclusive source of empirical evidence for political behavior and to use statistical techniques to analyze the information. Designed to explain the outcome of elections as much as to account for electoral decisions, the surveys emphasize factors that might vary systematically at the national level during election campaigns. Just as important, those who design the first national surveys deny the reliability of respondents' reports about the political preferences and behavior of their social intimates. Furthermore, the available statistical techniques could apply only to respondents who are independent of each other, a principle violated when members of social circles are included in the same survey. As a result, these surveys ask almost no questions that might provide direct information on the social context of politics.

Issues of theory offer another set of explanations for the turn away from the social logic of politics. The political scientists exaggerate the social determinism in the work of Lazarsfeld and his colleagues and then reject the distorted image. They insist that the electoral sociologists could neither explain electoral decisions nor the outcome of elections. Key and Downs deny what they see as the approach's presentation of nonrational or even irrational voters. As a result, the political scientists move to the analytical foreground the immediate determinants of vote choice: attitudes and calculations.

---

4 There are three exceptions to this research decision. In the first two national surveys of presidential elections (1956 and 1960), respondents are asked about potential personal sources of influence on their electoral decisions. In all surveys, respondents are asked questions about the objective social characteristics of members of their household and the political preferences of the respondents' parents.

5 Recent British Election Studies, however, provide information on political discussants, even as the American National Election Studies generally refrain from including these questions (see for example Johnston 1999; Pattie and Johnston 2000; 2001; Zuckerman, Kotler-Berkowitz, and Swaine 1998).

Social Logic of Partisanship

Once again, we sustain these points with statements taken from the critical books and articles. Consider first arguments about the effects of using national sample surveys, the absence of direct responses by social intimates, and the assumptions of the statistical techniques applied to the data:

> The structure of small groups has been successfully investigated by sociometric techniques but sociometry is difficult, if not impossible, to apply to large systems like nations. The macro-study of individuals was greatly aided by the development of the sample survey technique. But those who make most use of it – sociologists and social psychologists – are more interested in the behavior of individuals as individuals than in the structure and functioning of those large systems in which the political scientist is interested. (Eulau 1962, 134–5)

> [D]ifficulties in securing pertinent data have obstructed research on politically salient *dyadic* interactions for a sociological (rather than a psychological) understanding of individual on individual (interpersonal rather than group-on-individual) political effects. (Eulau 1986, 516)

The assumptions of most statistical studies provide another reason for the shift away from the analysis of the immediate social context of politics. Here, the point comes from a recent general criticism of research in social psychology:

> Dyadic relationships form the core element of our social lives. They also form the core unit of study by relationship researchers. Then why (to paraphrase Woody Allen) do so many analyses in this area focus on only one consenting adult at a time? The reason, we suspect, has to do with the rather austere authority figures of our early professional development: statistics professors who conveyed the cherished assumption of independent sampling.... How do we capture the psychology of interdependence with the statistics of independence?
>
> Unfortunately for the development of interpersonal relations theory, the patterns laid down by the imprinting period of graduate statistics classes tend to dominate the rest of one's professional life. Interdependence in one's data is typically viewed as a nuisance and so dyadic researchers have developed strategies to sweep interdependence under the rug. (Gonzalez and Griffin 2000, 181–2)

The data and methods available to examine the evidence on electoral choice move the analysis of partisanship away from the principles of the social context of politics.

The shift derives as well from theoretical considerations. The Michigan School reformulates, and the rationalists transform the explanatory mechanisms in the analysis of political preferences. Together they move the study of political behavior in general and partisanship in particular away from the social logic of politics.

Partisan Families

Consider first the flow of the argument in *The American Voter*. In the opening chapter, the authors justify the move away from community studies to a survey of the national electorate:

> In one important respect the research in Erie County and Elmira is only a partial account of the behavior of the American voter. Each of these studies has examined voting behavior within a single community. (Campbell et al. 1960, 15)

The volume then describes the National Opinion Research Center's first nationwide study of this sort, a survey of the 1944 presidential election:

> This study was prompted at least in part by the desire to extend beyond the bounds of a single community some of the generalizations suggested by the study of Erie County. (Ibid, 15–16)

Next, the authors detail the national surveys conducted for the elections of 1948 and 1952:

> The project represented a shift in emphasis from explanation in sociological terms to the exploration of political attitudes that orient the individual voter's behavior in an immediate sense. (Ibid, 16)

More fundamentally, they seek to link the analysis of electoral choice to the outcome of particular elections:

> This approach differed sharply from earlier sociological explanations and was intended to remedy some of the weaker aspects of these explanations. For example, the distribution of social characteristics in a population varies but slowly over a period of time. Yet crucial fluctuations in the national vote occur from election to election.... The attitudinal approach directed more attention to political objects of orientation, such as the candidates and issues, which do shift in the short term. (Ibid, 17; and see p. 65, where the authors cite Key and Munger 1959 on this point.)

Attitudinal variables stand close to the vote in the authors' "funnel of causality" (Campbell et al. 1960, 24–32). They merit, therefore, analytical priority.

Campbell et al. also maintain that it is not appropriate to use respondents' reports to characterize the political views and behavior of the members of their social circles (Ibid, 76). Emphasizing one of Festinger's principles (and ignoring others) and noting a point, which Converse would elaborate in later work (Newcomb, Turner, and Converse 1964, 126), the authors fear that people impute their own political views to the members of their immediate social circle. Hence, these data are tainted by problems of unreliability. Only questions relevant to the study of political socialization and descriptions of the objective social characteristics escape this decision.

Social Logic of Partisanship

Campbell, Converse, Miller, and Stokes do not, however, break completely with the social logic of politics. Even as they deny the reliability of respondents' reports, they affirm the theoretical importance of these perceptions:

> Yet this difficulty does not lessen our qualitative sense of the importance of the small group setting of partisan attitude and the partisan choice. And it does not obscure the finding from an analysis of errors of prediction that primary group associations may in the exceptional case introduce forces in the individual's psychological field that are of sufficient strength to produce behavior that contradicts his evaluations of political objects. (Campbell et al. 1960, 76–7)

Alas, they maintain, data problems inhibit their ability to follow their theoretical preferences.

The authors of *The American Voter* offer an ingenious – if flawed – solution to the conflict between the recognized need to include information on social contexts in the face of inadequate data. They alter the definition of the social group, conceptualizing it according to a person's perceptions and referring to a skewed interpretation of Kurt Lewin's ideas for support:

> [T]he distinctive behavior of group members was too obvious to leave unanalyzed. After a time the psychologist, Kurt Lewin, suggested a convincing resolution to the problem of the "group mind." "Groups are real," he said, "if they have real effects." Groups are real because they are *psychologically* real, and thereby affect the way in which we behave....
>
> Groups have influence, then, because we tend to think of them as wholes, and come to respond positively or negatively to them in that form.... Groups can become reference points for the formation of attitudes and decisions about behavior; we speak of them as *positive* or *negative reference groups*.
> (Ibid, 296)

And they also associate their position with Merton's scholarship, by referring to Norman Kaplan's Ph.D. dissertation *Reference Group Theory and Voting Behavior*, which was written under Merton's direction (and cited in both Merton 1957, 284, 332, 337 and Campbell et al. 1960, 297).

In their interpretation of Lewin's analysis, Campbell, Converse, Miller, and Stokes define groups not by patterns of interdependence or interaction but by a shared perception of a reference object, moving away from Lewin's definition of groups.[6] They also depart from Merton's understanding of reference group theory (1957, 237–386). Unlike Merton (and Kaplan's dissertation as described by Merton 1957, 284), *The American Voter* does not distinguish among "groups," which require the social interaction of members; "collectivities," which entail a sense of solidarity, shared values, and an "attendant sense of moral obligations to fulfill role

---

6 They do not offer a citation from Lewin's work in support of their solution.

expectations;" and "social categories," which he defines as "aggregates of social statuses, the occupants of which are not in social interaction" (Ibid, 299). Leaving Lewin's and Merton's path (as well as that of Lazarsfeld and the other electoral sociologists), Campbell, Converse, Miller, and Stokes emphasize the individual's perception of abstract referents, not the interactions of persons with each other, a flawed decision that is not supported by their own theoretical referents.[7]

As a result, they define partisanship as a psychological identification with (Campbell et al. 1960, 121) or attachment to (Ibid, 122) a political party. Strength of party support refers not to actions as much as feelings of intensity with regard to the reference group (and see, for example, Ibid, 122 for the initial formulation of the measure). In this conceptualization, people vary as well in the extent to which they identify with particular groups:

[T]he concept of group identification and psychological membership remains extremely valuable. Individuals, all of whom are nominal group members, vary in *degree* of membership, in a psychological sense. (Ibid, 297)

Let us think of the group as a psychological reality that exerts greater or lesser attractive force upon its members. (Ibid, 306)

And so, Campbell, Converse, Miller, and Stokes construct a survey question to assess how close the respondent feels to members of a group in order to measure group identification. This leads to their guiding hypothesis on the relationship between social context and political preferences:

[T]he higher the identification of the individual with the group, the higher is the probability that he will think and behave in ways which distinguish members of the group from non-members. (Campbell et al. 1960, 307; emphasis in original)

The conceptualization implies a set of related concepts and their measures. Cohesive groups have intensely loyal members (Ibid, 309), and the extent to which members "feel set apart" from others defines cohesiveness (Ibid, 310). "The political party may be treated, then, as a special case of a more general group-influence phenomenon" (Ibid, 331). Partisanship is psychological identification with a political party.

The American Voter offers two measures of social context: objective indicators such as education and occupation, and the subjective measures of identifications and feelings of strength or closeness. Except for answers to questions about the respondent's parents, it offers no information about

---

[7] This affects all subsequent work on party identification; see especially Miller and Shanks (1996); Green, Palmquist, and Schickler (2002).

Social Logic of Partisanship

the voters' immediate social circles. Subjective perceptions of social and political objects and reference groups replace patterns of trust, interdependence, and interaction among members of primary groups. Again, it is instructive to note how far this sense of group has come from the understanding offered by Lewin, Festinger, Merton, and Lazarsfeld and his colleagues, and their sources in the classics of social science. Again, it is important to underline that these decisions structure subsequent research on political behavior.[8]

V. O. Key and Anthony Downs present a more dramatic break with the social logic of politics. Both derive political choices from reasoned calculations about political objects. Key ignores the importance of members of social circles on political preferences, and Downs suggests that they play only a limited role. Both take giant leaps of theory.

Consider first how Key's position develops. As we note earlier, Key and Munger begin their article by criticizing *The People's Choice*. This volume, they fear,

> threatens to take the politics out of the study of electoral behavior. The theoretical heart of *The People's Choice* rests on the contention that "social characteristics determine political preference.... [Even though Lazarsfeld qualifies the statement,] [t]he focus of analysis...comes to rest broadly on the capacity of the "nonpolitical group" to induce conformity to its political standards by the individual voter....
> The study of electoral behavior then becomes only a special case of the more general problem of group inducement of individual behavior in accord with group norms. As such it does not invariably throw much light on the broad nature of the electoral decision in the sense of decisions by the electorate as a whole. (Key and Munger 1959, 281–2)

> A major burden of the argument has been that the isolation of the electorate from the total governing process and its subjection to microscopic analysis tends to make electoral study a nonpolitical endeavor.... Hence, all studies of so-called "political behavior" do not add impressively to our comprehension of the awesome process by which the community or nation makes decisions at the ballot box. (Ibid, 297)

V. O. Key's final work, *The Responsible Electorate* (1966, 7–8), extends the criticism beyond the research of Lazarsfeld and his colleagues to include *The American Voter* as well. Even as Key accepts the findings of Lazarsfeld's research (see earlier discussion, as well as numerous

---

8 Most of the chapters in the recently published collection of essays on political psychology (Sears, Huddy, and Jervis 2003) continue to ignore the social logic of politics. Even those that come closest (Huddy 2003; Lau 2003) do not stray far from the isolated individual whose social and political identities, perceptions, judgments, and actions are not much affected by members of their immediate social circles.

generalizations in the work with Munger 1959 and in *Public Opinion and American Democracy* 1961), he rejects the conceptual and theoretical implications for the understanding of the democratic citizen.

In *An Economic Theory of Democracy* (1957), Anthony Downs also breaks with the social logic of politics. Offering a perspective that defines rationality solely in political or economic terms, he maintains that members of intimate social circles provide no more than time-saving sources of information to calculating citizens. The effects of the immediate social contexts on people's preferences are well known, he concedes, but they stand in the way of a rational choice analysis of political behavior.

Downs begins his analysis by insisting that not every decision may be defined as rational. He examines, therefore, only the economic and political goals of persons and groups:

Admittedly, separation of these goals from the many others which men pursue is quite arbitrary.... Nevertheless, this study is a study of economic and political rationality, not of psychology....

Our approach to elections illustrates how this narrow definition of rationality works. The political function of elections in a democracy, we assume, is to select a government. Therefore rational behavior in connection with elections is behavior oriented toward this end and no other. Let us assume a certain man prefers Party A for political reasons, but his wife has a tantrum whenever he fails to vote for Party B. It is perfectly rational *personally* for this man to vote for Party B if preventing his wife's tantrums is more important to him than having A win instead of B. Nevertheless, in our model such behavior is considered irrational because it employs a political device for a nonpolitical purpose.

...Thus we do not take into consideration the whole personality of each individual when we discuss what behavior is rational for him.... Rather we borrow from traditional economic theory the idea of the rational consumer.... [O]ur *homo politicus* is the "average man" in the electorate, the "rational citizen" of our model democracy. (Downs 1957, 7; emphasis in original)

Undoubtedly, the fact that our model world is inhabited by such artificial men limits the comparability of behavior to behavior in the real world. In the latter, some men *do* cast votes to please their wives – and vice versa – rather than to express their political preferences. And such behavior is highly rational in terms of the domestic situations in which it occurs. Empirical studies are almost unanimous in their conclusion that adjustment in primary groups is far more crucial to nearly every individual than more remote considerations of economic or political welfare [where he cites Katz and Lazarsfeld 1955].

Nevertheless, we must assume that men orient their behavior chiefly toward the latter in our world; otherwise all analysis of either economics or politics turns into a mere adjunct of primary-group sociology. (Downs 1957, 8)

Downs offers a theoretical postulate in order to reject the well-founded observation that husbands and wives influence each other's political preferences. Claims for theoretical payoff justify this move, although he also offers an empirical observation (Ibid, 8).

Social Logic of Partisanship

Note as well that Downs – like Key and Campbell, Converse, Miller, and Stokes – retains a place for the principles of the social logic of politics. Toward the end of the book, he addresses the issue of the relationship between rational citizens and the costs of information. As citizens economize on time, they delegate the accumulation of information to others. Indeed, Downs maintains, people obtain information from others who share their views, again citing the work of Katz and Lazarsfeld and Lazarsfeld, Berelson and Gaudet (Ibid, 228–9).

Even as the behavioral revolution in political science begins by demonstrating the power of the primary group to explain partisanship and other political choices and action, it sets off in another direction. Even though the Michigan School takes incremental steps and Key and Downs bound away on leaps of theory, they reach the same point. Subsequent research pays little attention to dyadic relations; other intimate social circles, or workplaces and neighborhoods. Isolated respondents aggregated into nationally representative sample surveys provide the locus of study. Attitudes about candidates, policies, and issues proximate to the vote obtain theoretical primacy. Calculations about self-interest frequently tied to the principles of economic rationality predominate. The explanatory principles of the social logic of politics recede into the analytical distance.

RETURNING TO THE SOCIAL LOGIC OF PARTISANSHIP

Why return to the social logic of partisanship? Beyond the obvious and intuitive appeal, there are several reasons. As noted, even four decades ago when the reasons to reject it are at their best, they are not particularly strong. Even if one accepts that they were once appropriate, the problems that they addressed no longer stand, and the solutions are no longer necessary. As we show in this volume, surveys now offer representative national samples while also providing direct information on personal networks. As we elaborate in the empirical chapters, statistical techniques now permit analyses that detail the reciprocal effects of persons who live in the same households and take part in discussion networks. Developments in social psychology, decision theory, and bounded rationality move research past *The American Voter's* interpretations of Lewin and reference group theory and Downs's and Key's understandings of economic and political rationality. Also, as we show in this volume, when political scientists apply the social logic of politics, they are better able to account for partisanship and other political choices and action than when they do not. As a conclusion to this chapter, we develop each of these points.

There are now several sets of surveys that enable researchers to test hypotheses drawn from the social logic of politics on samples that represent national populations and electorates. One group, studies of political socialization, is a direct off-shoot of the Michigan School's interest in the household sources of partisanship (see, for example, Jennings and Niemi 1981; Stoker and Jennings 2005; Verba, Schlozman, and Burns 2005). The Cross National Election Survey applies Huckfeldt and Sprague's (1995) classic work on political discussion and communication to national samples (see, for example, Beck et al. 2002; Huckfeldt et al. 1995; Levine 2005). The Social Capital Benchmark Study 2000 provides data on the social context of a large sample of Americans (see for example Kotler-Berkowitz 2005), and British Election Surveys frequently include information on discussant networks. And clearly GSOEP and BHPS, the surveys that provide the evidentiary basis of our study, combine details on households into representative national samples.[9] Many of these surveys obtain information from all persons in the social networks, obviating concerns about the unreliability of data obtained from the reports of only one respondent. There is no longer a need to choose between Elmira, New York, or Erie, Pennsylvania, and the American electorate.

New statistical techniques help to detail the political interactions within households and among discussion partners. Structural and simultaneous equation models detail the relative importance of husbands and wives, household partners, parents and children, and similar sets of persons, on political choices (see for example, Stoker and Jennings 2005). Hierarchical models assess the relative significance of variables taken from the personal, household, neighborhood, region, and national levels of analysis (see, for example, Anderson and Paskeviciute 2005; Gimpel and Lay 2005; Johnston and Pattie 2005). In Chapters 4 through 6, we apply two- and three-stage models with instrumental variables in order to analyze reciprocal influence within households. We also apply techniques that adjust the standard errors, taking account of the presence of responses by persons in the same household and autocorrelation in times-series data. Principal–Agent models allow scholars to examine systematically and with theoretical rigor the tangled webs of social and political interactions (for example Huckfeldt, Johnson, and Sprague 2004; Johnson and Huckfeldt 2005). Interdependence among respondents no longer stands in the way of the application of powerful statistical techniques. There is no longer reason to study one consenting adult at a time.

To return to the opening section of this chapter, as social psychology, decision theory, and sociology – the initial sources of the behavioral revolution in political science – evolve, these disciplines continue to elaborate theories that underpin the social logic of politics. Even a cursory examination of studies on close relationships (Aron and Aron 2000; Cunningham

---

9 In a private conversation, Elihu Katz told Zuckerman that Lazarsfeld and his colleagues had planned such a survey, before they moved in other research directions.

and Barbee 1996; Haslam, McCarty, and Turner 1996; Hendrick and Hendrick 2000; Ickes and Duck 2000; Nye and Brower 1996; Oyserman and Packer 1996), the heuristics of bounded rationality (Arrow 1986; Gigerenzer and Selten 2001a, especially Gigerenzer and Selten 2001b; Jones 2001; Simon 1999), decision theory (Axelrod 1997a; 1997b; March 1994; Shafir and LeBeouf 2002; Shafir and Tversky 1995; Simon 1999), and social networks (Granovetter 1973; 1974) reaffirm the core principles: individuals learn from and influence each other; they create identities from the available social forms and the expectations of people around them; the boundary between the individual and the group blurs. As we noted earlier, anthropological theory agrees (Douglas and Ney 1998), and even developments in Freudian theory develop these principles (Alford 1994; Bion 1961). Even as its intellectual roots lay in the Bible and with the Greeks and the classics of modern social science, the social logic of politics also meshes with current theories that place individuals in their immediate social contexts.

At the same time, there is no reason to accept Key's and Down's separation of rational choice theory and the social logic of politics (Huckfeldt and Sprague 1995; Huckfeldt, Johnson, and Sprague 2004; 2005; Johnson and Huckfeldt 2005). There is also every reason to take account of fundamental critiques of rational choice theory, which challenge its understanding of how decisions are made (Gigerenzer and Selten 2001a; 2001b; Jones 2001; March 1994; Shafir and LeBeouf 2002; Shafir and Tversky 1995; Simon 1999) and its atomistic assumptions (Douglas and Ney 1998; Gigerenzer and Selten 2001a; 2001b; March 1994). In the concluding chapter, we suggest that there is a reciprocal relationship between the partisan influence within households and competition among the political parties. Campbell, Converse, Miller, Stokes, Key, and Downs imposed a false choice on the study of partisanship.

Furthermore, applying the social logic of politics helps to resolve puzzles associated with rational choice approaches to political behavior. Using a "small world" logic of turnout, Fowler (2005) demonstrates that decisions to go to the polls influence the probability that others in the social network will also vote, thereby, increasing the incentive to cast a ballot. Mebane (2004) shows that cue-giving helps to bridge the gap between evidence that voters are engaged in large-scale strategic coordination and evidence that most voters are too disengaged from politics to do anything like that. In turn, Huckfeldt, Johnson, and Sprague (2004) and Johnson and Huckfeldt (2005) explore the bases of opinion diversity in the face of logics that expect opinion conformity. Furthermore, they provide an explanatory mechanism that is both simpler and more accurate than Achen's (2002) complex claims about political and other learning in households. These are but a few examples of the successful application of the social logic of politics to areas heretofore examined by rational choice theorists.

At least as important, scholars who apply the social logic of politics display analytical successes that move scholarship beyond the expectations of the Michigan School and rational choice theory. That, after all, is the fundamental claim of this book. Even relatively distant friends influence each other, and almost everyone's political views and behavior are affected by the views of others in their neighborhoods (see, for example, Gimpel and Lay 2005; Johnston and Pattie 2005; Kotler-Berkowitz 2005; Levine 2005; Lin 2005). As we show, household partners and members of discussion networks influence each other's partisan preferences, no matter their individual-level characteristics (see also Stoker and Jennings 2005; Zuckerman and Kotler-Berkowitz 1998). Social ties and interactions – not only perceptions of reference objects – influence political choices and behavior.

Finally, we highlight the relationship between our application of the social logic of politics to the dominant schools of research on partisanship – the Michigan School's emphasis on party identification (in addition to the foundational work examined earlier, see especially Converse 1969; 1976; Green, Palmquist, and Schickler 2002; Miller and Shanks 1996) and classical rational choice theory's image of partisanship that derives from personal calculations. On conceptual and theoretical grounds, we reject the Michigan School's equation of psychological and social membership in groups. On these grounds too, we deny the claim that partisanship is necessarily stable – immune to other forces. Indeed, GSOEP and BHPS evidence shows three important patterns as people respond to questions about partisanship over time: they hardly ever move from party to party, but they almost never support the same party again and again, moving instead between naming a party and announcing no partisan preference. Because most people never support one or the other of the major parties and because they stay on the same side of the political divide, correlations of their responses over time are necessarily high (see especially Green, Palmquist, and Schickler 2002). Our analysis shows that partisanship is something other than a fundamental loyalty that only strengthens with age. At the same time, when Green, Palmquist, and Schickler observe, "partisan stability is traceable to constancy in citizens' primary group environment," as well as to psychological factors (2002, 21), we are heartened to find an effort to recover the Michigan School's reliance on the social logic of politics.

The social logic of partisanship joins easily with subjective or bounded rationality, if not efforts to apply rational choice theory to partisanship (see, for example, Achen 1992; 2002; Downs 1957; Fiorina 1981; 2002; Kiewiet 1983; Shively 1977). Towards the end of *An Economic Theory of Democracy*, Downs addresses the relationship between rational citizens and the costs of information. As citizens economize on time, they delegate the accumulation of information to others. Indeed, he maintains, people look to obtain information from others who share their views, and approvingly cites the Columbia School (Downs 1957, 228–9). Developing this connection, Huckfeldt and Sprague's research (for example 1995; as well as Fowler 2005; Huckfeldt, Johnson, and Sprague 2004; Mebane 2004) on political networks provides a bridge between Downs and Lazarsfeld and his colleagues. Learning from others, from our perspective, is not solely a matter of saving time; it is also what humans do *qua* humans.

And so, we apply a series of related hypotheses to the empirical portions of our work: In the two established democracies that we study, social influence on partisanship is always probabilistic, never determined. Each person may opt to ignore the preferences of others; most do not. The influence of members of intimate social circles (families and households) affects the elements of partisanship and electoral choice, net of all other explanatory variables. Indeed, the power of these social interactions stands out among all sources of partisanship and electoral behavior. Within social groups political influence is usually reciprocal, not one-way. The political cohesion of any social group – even the most intimate – is a variable. The more that people interact and the more they share, the more likely are they to influence each other.

2

Bounded Partisanship in Germany and Britain

Partisanship entails a series of related choices. At two analytically distinct but related moments, people decide whether to support a political party or not (*partisan support*) and which party to name (*partisan preference* or *choice*). In Germany, most people say that they incline towards a political party, either the Social Democrats or the Christian Democrats/Socials, and in Britain, most report that they support either Labour or the Conservatives. At a single point in time, selecting a party entails its complement: not naming another party; these are mutually exclusive choices. In both countries, these are rather straightforward outcomes.

Over time, however, choosing between the two parties is more complex. Always naming Party A implies never choosing Party B or any other party, but never choosing B does not entail always naming A or any other party. Sometimes choosing A also implies nothing about the selection of other parties: persons vary in the extent to which they name one of the political parties, and they vary too in the extent to which they move between naming a party and announcing that they prefer no party. Observed over an extended period of time, *partisan constancy* (the rate of partisan preference) in both Germany and Britain displays large numbers of persons who never support Party A/B and variation along a scale of frequency of choice for Party B/A. Very few move between A and B. Most people behave as if they construct a choice set from the competing political parties, excluding one of the major parties from the frame and then deciding each time whether or not to prefer its major rival. As a result, partisanship is a limited – or bounded – choice.[1]

---

[1] See Markman and Medin (2002) and the literature that they cite on the creation of choice sets. Bannon (2003) provides a logic for the segmentation of markets (i.e., excluding some and focusing on others), which applies to businesses in search of consumers and political parties in search of voters.

Bounded Partisanship in Germany and Britain

As we see it, the analytical choice is not between persons who repeatedly name the same party (as implied by the Michigan School of electoral analysis; see especially Campbell et al. 1960; Green, Palmquist, and Schickler 2002; Miller and Shanks 1996) and those who evaluate their own and the general circumstances and match them against the parties' past actions and future promises (as supposed by those who apply classical rational choice theory to partisanship; see, for example, Achen 1992; 2002; Fiorina 1981; 2002; Kiewiet 1983; Shively 1977). Neither describes the persons whom we observe in our analyses. Rather, most people delimit the objects of selection to one of the major parties and then decide whether to name that party at that point in time and again and again.

In this chapter, we detail partisan support, preference, and constancy in Germany and Britain. As we note in the Preface, we take our data from two nationally representative household panel surveys: the German Socio-Economic Panel Study, which describes Germans during nearly two decades (1985–2001), and the British Household Panel Survey, which examines people in Great Britain during the years 1991–2001. In subsequent chapters, we apply hypotheses drawn from the social logic of bounded partisanship to account for these patterns.

PARTISANSHIP AT ANY GIVEN POINT IN TIME

This is a simple story. Taking all the answers to the questions that tap partisan support and preference by persons who are ever in the two surveys, we present a picture of these two dimensions of partisanship in an "average year." In Germany, the many responses to GSOEP show that 0.25 of the sample pick the Social Democrats, 0.22 name the Christian Democrats/Socials, 0.44 select no party, and 0.09 name some other party. The data from BHPS report the following breakdown: Labour, 0.32; Conservatives (Tories), 0.22; no party, 0.34; and other parties, 0.11. Here, the analytical task is to account for the probability of partisan support and the probability of naming each of the dominant parties. In the next chapter, we analyze these choices, in a two-stage model that relates partisan support and preference and that demonstrates the explanatory power of members of the household on these decisions.

First, we elaborate partisan constancy – the rate by which a party is named (chosen or selected) over time. This enables us to provide a more complex picture of partisanship, detailing the presence of bounded partisans. We offer these patterns both at the level of the citizenry, the people taken as a collective, and citizens, focusing on individual decisions about the political parties, over time.

Partisan Families

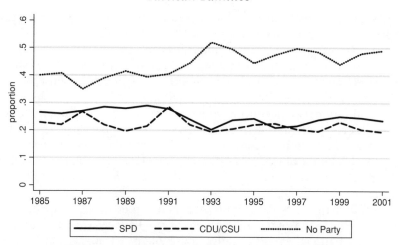

Figure 2.1. Aggregate Party Preferences in Germany

THE DYNAMICS OF PARTISANSHIP

As individuals make decisions about political parties, so do members of their family, their friends, neighbors, and other fellow citizens. This obvious point has three important implications for our analysis. (1) As we develop in this volume, each person's choice is influenced by and influences those made by their social intimates and others with more distant social connections. We return to this theme in the next chapter; here, we emphasize two other consequences of partisanship as a social relationship. (2) Brought together, individual decisions form aggregate patterns. The citizenry (i.e., citizens taken as a collectivity) chooses between the two major parties. What seems to be a real choice by the citizenry is a limited decision for individual citizens.[2] (3) Furthermore, there is an interactive relationship between the aggregate distribution of partisan preference and the relative size of the parties. This point, no matter how obvious, needs to be affirmed. Large parties are highly visible and have the resources to mobilize support, and for these reasons alone people are prone to name them. We return to the reciprocal relationship between the parties' efforts to mobilize support and people's decisions about the parties in the final chapter. In this section, we examine aggregate temporal support for political parties among the citizenry, the citizens as an aggregate.

At any point in time and over time, the citizenry is more or less equally likely to name one or the other of the major parties. Figures 2.1–2.4

---

[2] Taber (2003) distinguishes between the two levels, and see Erikson, MacKuen, and Stimson (2002) and Page and Shapiro (1992) on rational publics.

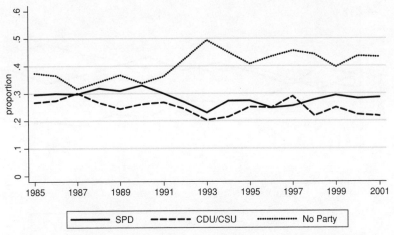

Figure 2.2. Aggregate Party Preferences among Germans in All Waves

highlight this general claim. From this perspective, partisanship is a choice between two political options. This aggregated competition stands as the fundamental basis of democratic rule; each set of German and British party leaders and expectant government leaders knows how close they are to being voted out of or into office (Sartori 1976).

How do the aggregate trends speak to our understanding of macropartisanship in each and both countries? Do we find evidence of secular declines? Do we see growth in the level of partisanship as persons grow older during the years of the surveys? Do we find evidence that partisanship responds to political events? These questions introduce classic themes in the scholarship on partisanship.

Examining Germany over seventeen years, Figure 2.1 highlights growth and decline in the aggregate level of partisan preferences, offering a picture that agrees with numerous observations of recent German politics (for example Dalton and Bürklin 2003). Here, we present the responses of each cross section over time, describing the partisan distribution of persons interviewed in that year. Figure 2.2 displays the responses of the smaller sample: persons interviewed in all waves, during the years 1985–2001.[3] Note the similarity between the figures; the only apparent difference lies in the somewhat greater tendency of those in all years of the panel

---

3 The parallel results reaffirm the validity of claiming that GSOEP's panel survey is a representative sample of the German population, even though it is populated by persons who are more likely to live in the same place than the cross-sectional samples. The same results characterize the measure of political interest that we describe later, again reaffirming the utility of the panel data.

Partisan Families

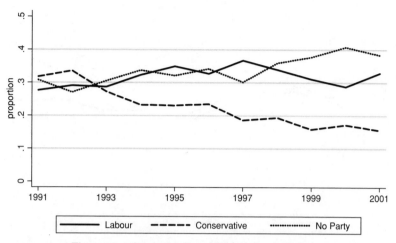

Figure 2.3. Aggregate Party Preferences in Britain

to support a party in each year and over time. During the first six years, the trend lines follow the expectations of the classic literature on partisanship (Converse 1969; 1976). Veteran citizens display a relatively high and stable level of partisan preference. This familiar trend stops at Reunification, as this watershed event is followed by a drop in the aggregate level of partisanship that stops only with the 1994 election, in both the cross-sectional and panel samples. Note however that during this period, partisanship does not spike in election years – 1987, 1990, 1994, and 1998 (but see Zelle 1998). Indeed, it is difficult to see any evidence that partisan choices respond to political events. During the data's nearly two decades, aggregate partisan preference declines, even as support for the two major parties does not change much.[4]

Figure 2.3 displays the aggregated results for persons who are interviewed at least once in the BHPS survey, and Figure 2.4 presents the same trends for persons in all waves. Echoing numerous surveys and the results of the two general elections, both depict a citizenry moving away from the Conservative Party with a concomitant rise in the percentage that declines to choose any party (though these trends are somewhat modulated among those in all waves). No dramatic shifts appear to parallel

---

[4] Because the GSOEP data begin in the mid 1980s, they do not enable us to address directly the observation that partisanship has steadily declined since the early 1970s (Schmitt and Holmberg 1995; Dalton and Wattenberg 2000, especially Table 2.1, p. 25). Our year-to-year evidence indicates short-term variations within a slight decline in the aggregate level of partisanship, but see the discussion that follows. See Kohler (2002) and Schmitt-Beck, Weick, and Christoph (2006) for related analyses of German partisanship.

Bounded Partisanship in Germany and Britain

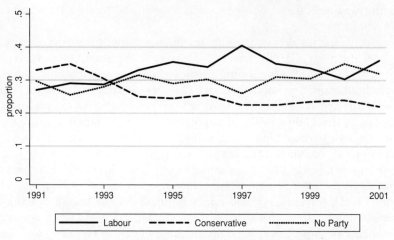

Figure 2.4. Aggregate Party Preferences among Britons in All Waves

Thatcher's departure and Blair's and Labour's ascendance. Note two primary points of transition: the years after the election of 1992 coincide with the start of the Tories' decline, and 1997 is the peak for Labour, as it also initiates a rise in the percentage that supports no party.

Trends in the citizenry's partisanship call attention to the secular decline in partisan preference that has characterized many established democracies. In Germany, the decline grows to characterize about half the people by the turn of the last century. In Britain, there is less a generalized drop in partisan support than a reduction of preference for the Conservatives. At the start of the GSOEP survey, approximately 0.40 of the Germans (in the cross-sectional surveys) claim to support no party, a level reached in Britain more than a decade later. These results also hint at a disjunction between the citizenry and the citizens. As persons remain in the panel surveys they necessarily grow older, but aging does not increase the probability that they name a party (see Figures 2.2 and 2.4). This is the case even though, as we show later in the volume, the positive relationship between age and partisanship remains in the cross section. Furthermore, powerful exogenous shocks to the political system seem to have little effect on the trends in aggregate partisan support and choice.

POLITICS IN THE DAILY LIVES OF GERMANS AND BRITONS

One reason that the events of the day hardly affect aggregate levels of partisanship is that politics is far from most people's concerns. In order to elaborate this picture, we supplement the GSOEP and BHPS surveys with responses taken from the second wave of the World Values Surveys (WVS) in 1990.[5]

Most Germans do not attribute much importance to politics. In 1992, 1994, and 1995, GSOEP asked people to rank the importance of a series of items in their lives, providing a scale, where 1 = unimportant, 2 = not very important, 3 = important, and 4 = very important. The mean score for all the items over the three years is 3.04. Political activity ranks lowest, by far. In descending order of importance, they are family (3.7), happy marriage or partnership (3.6), dwelling (3.44), environment (3.38), children (3.26), income (3.33), leisure (3.12), being there for others (3.10), work (3.00), being able to buy things (2.95), circle of friends (2.91), owning a home (2.76), being fulfilled (2.70), success at one's job (2.67), religion (2.52), travel (2.45), and finally political activity (1.98). WVS data repeat these patterns for Britain as well as Germany. Again in descending order of importance, we find: the family (3.65 in Germany and 3.84 in Britain), friends (Germany: 3.25; Britain 3.4), leisure (Germany: 3.22; Britain: 3.31), work (Germany: 3.09; Britain: 3.17), religion (Germany: 2.74; Britain: 2.58), and then politics (Germany: 2.37; Britain: 2.33). Of these dimensions of life, politics is the least important.

Relatively few take part in local political activities. At eight opportunities, GSOEP asks people about their level of participation in local politics, a relatively easy mode of political activity. At seven occasions, about 0.90 reply that they do not take part at all; the exception again is 1991, when 0.20 claim to participate at least sometimes in local politics (and for similar results, see Parry, Moyser, and Day 1992, 44; Pattie, Seyd, and Whiteley 2004; Kaase 1989; van Deth 1989; Verba, Schlozman, and Brady 1995, 70). Again and without much surprise, WVS reports that less than 0.03 of the Germans and not even 0.02 of the Britons are active members of political parties (a portion not much different than active members of other formal voluntary organizations). Hardly any persons are active members of social or political organizations of any kind.

The level of interest in politics is also consistently low, no matter that the Germans witness the transformation of their polity and the Britons live through an electoral earthquake. GSOEP probes this issue in every wave and BHPS asks the respondents about their interest in politics in the first six waves and returns the question in 2001. Figure 2.5 displays the levels of political interest in Germany. The responses vary between 2.23 and 2.44, and the peak – in 1991 – is a spike associated with German Reunification

---

[5] These are representative samples of the respective populations. We obtained the data from the Inter-University Consortium for Political and Social Research, and we thank the consortium and Ronald Inglehart, who directs the World Values Surveys.

Bounded Partisanship in Germany and Britain

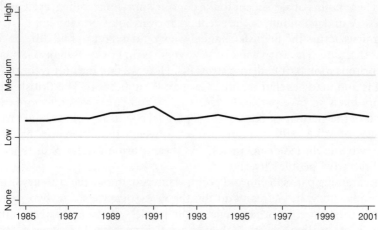

Figure 2.5. Aggregate Levels of Political Interest in Germany

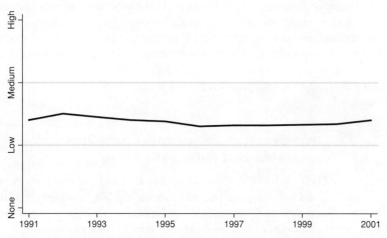

Figure 2.6. Aggregate Levels of Political Interest in Britain

(see Figure 2.5). Note that the substantive difference between these two scores on the 300-point scale (1.00–4.00) is 0.07 – not much at all! In Britain, the aggregate results also do not much vary, ranging between 2.25 and 2.47, with the highest score in the year of the 1992 general election (see Figure 2.6). Note too that this is the same substantive difference. Again WVS reports similar results, with a mean in 1990 of 2.87 in Germany and 2.43 in Britain. Political parties stand far from most people's lives, even during Germany's years of political turmoil and electoral transformation in both Germany and Britain.

These points of agreement to the contrary notwithstanding, even with a cursory understanding of the events that overtook Germany but have no parallels across the English Channel we would expect to find differences, and there are. The 2000 wave of WVS reports that twice as many Germans as Britons took part in demonstrations. Here we find the only trace of the vigils and marches that marked German Reunification. The British were more likely to take part in boycotts (0.17 compared to a bit more than 0.10 of the Germans). Both of these parallel results reported in the earlier waves of WVS. Still, the similarities outweigh the differences between the two established democracies, and these affirm the relative distance of politics from people's lives.

Examining partisanship and political interest among the citizenry offers a simple story. In Germany during the years that we study, two parties battle toe-to-toe. In Britain, one of the parties, the Conservatives, takes a political beating, and the other, Labour, dominates. The citizenry is no more than mildly interested in their contests.

Observing citizens rather than the citizenry offers a more nuanced and theoretically rewarding story because it details complex patterns of partisan choice over time. In the next section, we show that individuals usually never name one of the major parties. Over time, they vary their preference for its major opponent, moving between selecting it and between claiming to prefer no party. They are bounded partisans. In the next chapter, we begin to account for the bounded nature of partisanship in Germany and Britain.

THE DYNAMICS OF MICROPARTISANSHIP: BOUNDED CHOICE OVER TIME

Because GSOEP and BHPS offer so many data points, we are able to focus on various meanings of partisan constancy. We consider the most direct interpretation of this concept: (1) the pattern of change in partisan choice between points in time and (2) the constancy of partisanship over many points of time (i.e., how frequently, or at what rate, do persons agree with themselves each time that they choose?). In order to address the first point, we aggregate all responses ever offered in each survey, and we compare answers between two adjacent years. This sample examines all persons who reply at least twice, and we present "average" responses. The second statement examines the persons who take part in all waves of the two surveys. In this case, we count the number of times that persons name the same political party during the designated period of time, and for ease of interpretation, we turn these results into percentages. Unlike correlations, count results do not describe whether individuals stay at the same place on a scale relative to all others in the sample, and they are not

Bounded Partisanship in Germany and Britain

Table 2.1. *Partisan Change between Any Two Adjacent Years*

	SPD at $T-1$	CDU/CSU at $T-1$	Other party at $T-1$	No party at $T-1$
SPD at T	0.78	0.02	0.10	0.13
CDU/CSU at T	0.02	0.76	0.14	0.11
Other party at T	0.03	0.04	0.47	0.05
No party at T	0.18	0.18	0.27	0.70
	Labour at $T-1$	Conservatives at $T-1$	Other party at $T-1$	No party at $T-1$
Labour at T	0.80	0.01	0.14	0.18
Conservatives at T	0.01	0.92	0.03	0.08
Other party at T	0.02	0.01	0.65	0.04
No party at T	0.16	0.06	0.18	0.70

sensitive to the mean and distribution of opinions. They simply detail the extent to which persons announce support for the same party again and again.[6] These two sets of answers structure much of our analysis in this chapter. A third technique also observes decisions over time, by placing them on a grid and noting the patterns. Here, we use columnar reports in order to display the paths taken in each sample over all the years of the survey. These offer complementary measures for partisan constancy. Taken together, they show that most persons never support Party A/B and vary their support for Party B/A; they constrain the choice set to that of naming and not naming one of the two dominant political parties.

Tables 2.1 and 2.2 distribute the partisan preference of the respondents in each survey at two adjacent average points in time. In Table 2.1, we show the partisan choice at T of respondents who selected one of the four options at $T-1$ (in Germany: SPD, CDU/CSU, another party, and no party; in Britain: Labour and the Conservatives as well as the other two choices). Simply put, Germans and Britons who pick one of the dominant parties usually name that party again. When they change, they usually name no party. Table 2.2 offers a global perspective on the electorates. Of

---

[6] For an elaboration of the distinction between agreement and reliability as measured by correlations, see the classic statement by Robinson (1957) as well as more recent analyses (e.g., Tinsley and Weiss 2000). Following the exhortations of Green, Palmquist and Schickler (2002), we also calculated the polychoric correlations between the responses on partisan choice across each set of years in the surveys, among persons in all waves. Because the pattern of actual choice is bounded, these correlations are always very high. We do not present the tables because they are of little theoretical interest and take up several pages, but they are available from the authors.

Partisan Families

Table 2.2. *The Electorates Distributed by Partisan Choice between Two Points in Time*

	SPD at $T-1$	CDU/CSU at $T-1$	Other party at $T-1$	No party at $T-1$
SPD at T	0.20	<.01	0.01	0.05
CDU/CSU at T	<0.01	0.17	0.01	0.04
Other party at T	<0.01	<0.01	0.05	0.02
No party at T	0.05	0.04	0.02	0.30
	Labour at $T-1$	Conservatives at $T-1$	Other party at $T-1$	No party at $T-1$
Labour at T	0.26	<0.01	<0.01	0.05
Conservatives at T	<0.01	0.18	<0.01	0.03
Other party at T	<0.01	<0.01	0.04	0.01
No party at T	0.04	0.03	0.01	0.24

the sixteen possible combinations, three predominate: repeated naming of a dominant party and no party. In Germany, these account for 0.70 of the responses, with naming no party twice the plurality choice; in Britain, these decisions amount to 0.60, with repeated picks of Labour the most frequent combination. No more than 0.01 move from one major party to the other in Germany and 0.02 do so in Britain. Note too that most of the partisan switching in adjacent years in each country takes persons between no party and one of the dominant parties. Partisan choice evidences highly constrained – or bounded – patterns.

Now we examine persons who respond to the partisanship questions in all waves of each panel. This is a direct measure of partisan constancy. Figures 2.7 and 2.8 present the strikingly similar patterns in both countries; there is a sharp spike at the zero count, a drop to relatively few selections at each of the count points, and then a slight rise at the end.[7] The only differences between the two countries reflect the higher proportion of Britons who consistently pick a party and the relative strength of Labour during the decade under analysis.[8] In both Germany and Britain, large numbers of citizens never support one or the other of the major parties and vary their support for their party of choice (and see especially

---

[7] The relative dearth of recurrent selections further highlights the internal validity of the data on party support. Repeated questions do not seem to produce the same answers; there is no evidence here of instrument effects in which the survey questions prompt the respondents' answers.

[8] These results closely resemble those found by Schickler and Green (1997) and Green, Palmquist, and Schickler (2002), who examined British and German panel evidence from earlier and shorter periods of time.

Bounded Partisanship in Germany and Britain

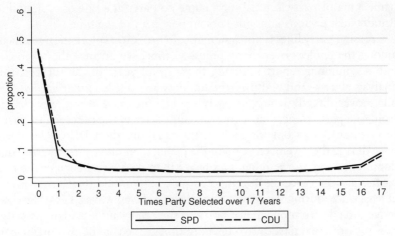

Figure 2.7. Partisan Constancy in Germany

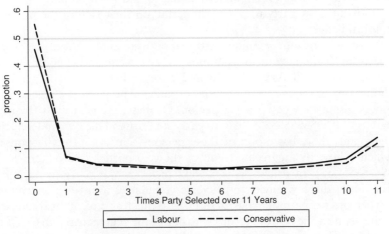

Figure 2.8. Partisan Constancy in Britain

Schmitt-Beck, Weick, and Christoph 2006 for a parallel analysis of Germany using GSOEP data).

Crossing party lines is infrequent. Germans who name the SPD at least once pick their party 0.53 of the time and the CDU/CSU 0.07. Those who ever select the CDU/CSU name their party 0.49 of the opportunities and the Social Democrats 0.09. Persons who choose Labour at least once during the eleven years of the survey pick Labour 0.62 of the time and the Tories in 0.06 of the opportunities. Those who ever choose the Conservatives name their party at a rate of 0.60 and Labour at 0.07. When they do not pick their party, they usually select no party. Germans who

43

Partisan Families

name a major party at least once, name no party at a rate of 0.33, and in Britain such persons announce no support 0.25 of the time.

Crossing party lines is not systematic. In order to explore the movement among the choices over many points in time, we arrange the responses into a columnar report.[9] Among those in all years of the surveys, this analysis shows (and reaffirms the findings in Table 2.1) that hardly anyone moves directly from one party to the other, and even fewer change from one party to become a constant supporter of the other major party. In Germany, 0.17 opt for each of the SPD and the CDU/CSU at least once during the seventeen years of the panel; 0.07 move directly from one dominant party to the other, and fewer than 0.03 select each of the two parties at least twice. In Britain, 0.12 pick Labour and the Conservatives at least once, during the years of the survey's eleven waves. Less than 0.04 move directly from one major party to the other and less than 0.04 also ever pick the two parties two or more times. Even as the two major parties experience profound leadership and policy changes, almost everyone alters their partisan preferences by going to and from no party, but not by moving from party to party. They stay on their side of the national political hedge.

Most persons pursue unique paths of partisan choice over time. Their selections distinguish them from most everyone else and from the aggregate trends as well. As we know from Figure 2.7 in Germany, 0.22 always choose one of the two major parties or indicate no party preference. The others display 1.3 patterns per person. Of these, the next most frequent pattern (NSSSSSSSSSSSSS, first no party support and the remainder Social Democrat) describes 0.007 of these persons. No one offers responses, which follow the aggregate trend of the two parties (the pattern SSC-SSSSSSSSCSS). In Britain, 0.31 always pick a major party or no party. The others display remarkably diverse preferences over time, on average 1.5 persons per pattern. No path receives as much as 0.01 of the responses. Consider now the case of persons who mimic the aggregate trend: CCCLLLLLLLL. These account for 0.001 of the sample! Aggregate trends are the result of individual decisions, but hardly anyone behaves like the national results: citizens' choices are not the same as the citizenry!

Now consider those persons who name each major party at least once. How do they choose over time? Among Germans, the mean rate of selection for the SPD is 0.29; for the CDU/CSU, 0.24; another party, 0.06;

---

[9] This technique displays the responses for each year as a column in a table that includes all years and all persons in the surveys. We labeled the parties as follows: in Germany, S = Social Democrats and C = Christian Democrats/Socials; in Britain, L = Labour and C = Conservatives in both O = Other and N = None. Interested readers may obtain the complete distributions from the authors.

Bounded Partisanship in Germany and Britain

and no party, 0.40. In Britain, it is Labour, 0.36; the Conservatives, 0.27; another party, 0.07; and no party, 0.36.[10] Note as well that the rate of cross-party movement over time equals the number of years in the panels (0.17 in seventeen years of GSOEP and 0.12 in BHPS's eleven years). These patterns fit the behavior of those persons who behave as if they are "flipping" partisan coins each year. They may also be in line with those whose careful calculations find significant distinctions between the two parties. Either way, it is important to recall that these are relatively small segments of each population. Most persons do not behave as if their partisan movement is haphazard or calculated. Rather, as we show in the following chapters, partisan support, choice, and constancy reflect the particular social circumstances of their lives.

In the concluding chapter, we return to this issue. There, we suggest that party mobilization interacts with household effects to bring national political events into people's partisan decisions and electoral choices. Because so few persons move directly from party to party or even ever name both of the parties over time, we do not analyze this problem (but see Kohler 2005).

SUMMARY, IMPLICATIONS, AND THE NEXT STEPS

There seems to be no relationship between partisanship at the aggregate level – partisan choice of the citizenry writ large – and the microdynamics that we have just observed; other than the former emerges from the latter. The diverse trend lines found in the columnar reports indicate that individuals over time do not make partisan decisions in concert, as if they are part of a collective. Analysis of aggregate partisanship obscures this point. As a result, we focus on the determinants of partisanship, as they affect the decisions of citizens, not the citizenry.

Examining the partisan choices of Germans (1985–2001) and Britons (1991–2001) shows that they are bounded partisans. Most stand apart from one or both of the major parties and vary between naming its rival and claiming to support no party. Hardly any ever select both of the major parties during the years of the surveys. And so, we suggest that people construct a very limited choice set from the full list of political parties.

The presence of bounded partisans raises questions about widely accepted approaches to partisanship in political science. Rational choice theorists need to account for the construction of a choice set in party systems with two large parties and the limited movement from party to

---

10 Kohler (2002) finds results very similar to those that we report in an analysis that looks at the Greens and the Free Democrats as well as the SPD and CDU/CSU and no party.

party over so many years. The approach of the Michigan School needs to puzzle over the relative dearth of people who persistently prefer the same party. Variation in partisan choice over time is not an illusion derived from random measurement error.[11] Partisan constancy is not a simple extrapolation of partisan choice. Furthermore in Chapter 6, we show that partisan constancy directly affects vote choice. Partisan constancy merits direct analysis; it may not be assumed to derive from partisan choice.

What accounts for patterns of partisan choice and constancy, which result in bounded partisanship in Germany and Britain? Most generally stated, we maintain, persons send and take cues from others with whom they interact, especially those with whom they live. The distribution of party preferences in the household influences each person's choice of party, at a single point in time and over time. Bounded partisans send cues to and receive cues from other bounded partisans.

In the next chapter, we focus on the decisions of heads of households, leaving a more comprehensive analysis to a later point in the study. We turn party support and preference into variables, seeking to account for the probability that the head of household will pick a party and name a particular one. Drawing on the social logic of politics and making use of the data from GSOEP and BHPS, we offer statistical models of partisan support and preference, which show the importance of family members on the decisions of the household's head, net of the effect of his or her other social characteristics, age, and other personal characteristics including political interest and economic concerns. We extend this analysis to the question of partisan constancy as well, and again display the explanatory importance of the distribution of partisan preferences on the probability of naming a party at an "average" moment and the rate of that choice over time.

---

[11] The classic sources for the debate over the stability of political attitudes are Achen (1975) and Converse (1964), but see Saris and Sniderman (2004) for a recent collection of essays on this theme.

3

A Multivariate Analysis of Partisan Support, Preference, and Constancy

Partisanship in Germany and Britain reflects the analytical ties among religion, social class, and party that have characterized European politics for decades and that persist, even as indications of "individualized politics" appear.[1] More fundamentally, however, our analyses underline the importance of immediate social and political networks as contexts for decisions about the political parties. Individuals choose political parties, at any point in time and over time, by taking into account the perceptions, values, actions, and cues of persons in their social networks. The stronger and the more frequent the social tie, the more powerful is the influence; families and households are especially important. There is a social logic to bounded partisanship.

MODELING PARTISANSHIP

In this chapter, we begin the analysis of partisan support, preference (choice), and constancy. Note that no one can select or choose a party without first claiming to support a party. Over time, the rate of picking a party requires naming or selecting it at least once. The initial survey question implies a simple answer: to support or not to support a party. In Germany and Britain, the second choice usually means selecting one

[1] Representative recent examples of this voluminous literature include the essays in Dalton and Wattenberg (2000); Evans (1999a); Franklin, Mackie, and Valen (1992) as well as Dalton (2000); Gluchowski and von Wilamowitz-Moellendorff (1998); Norpoth (1984); Richardson (1991); Schmitt (1998); Schmitt and Holmberg (1995); and Sinnott (1998). As we find strong class and religious effects on partisanship in both countries, we lend support to Evans (1999b; 1999c); Goldthorpe (1999a; 1999b); Kotler-Berkowitz (2001); Müller (1999); Weakliem and Heath (1999). Applied to Germany, see for example Dalton and Bürklin (2003); Falter, Schoen, and Caballero (2000); and Kohler (2005).

of the two dominant parties. We model each as a bifurcated decision: to support or not to support a party and to prefer or not Party A/B.

The connection between partisan support and party preference counsels that we apply the Heckman Probit Selection model[2] to analyze our survey data. This model employs a two-step statistical procedure that first calculates the *select*[3] component of the model (here, party support or not) and then uses this response to estimate the second (*outcome*) part of the equation – Party A/B or not. The results of our analyses in this chapter serve as the base-lines for all the other models in this volume. They are the most elaborate, including all the variables in the data set that our theory and its rivals suggest influence partisan support and preference.

Later in the chapter, we answer questions about partisan constancy. In that work, we apply a count model that addresses both the probability of ever picking a particular party and the frequency of coming back to that party over time. Because so few persons move from party to party (see Chapter 2), we do not analyze partisan switching. In order to estimate both the Heckman Probit Selection and count models, we use Stata/SE8.2.

In abstract form, we derive specific hypotheses from the social logic of politics to account for partisan support, preference, and constancy. These answers come in three forms. The first appears in the results of the models themselves. These show the impact of particular variables on the outcome variables, net of the effect of the other predictors. Second, we draw postestimation probabilities to simulate the effects of specified combinations of characteristics on the predicted probability on each of the dependent variables, and third, we compare the results to the sample means.

OVERCOMING ENDOGENEITY IN HOUSEHOLD DATA

As noted, we are particularly interested in how persons who live together influence each other's partisan choices. This expectation entails the need to clear a serious technical hurdle, before analyzing our data. If we expect person A to influence person B and B to influence A in turn, both at levels

---

[2] We use this model (rather than a more conventional Heckman Selection model) because each of the dependent variables (partisan support and choice) is dichotomous. This model also corrects for selection bias in the sample occasioned by excluding persons who claim no party support. Note that there are no widely used multinomial models equipped to handle the problem of selection bias. For more on the model, see the Appendix.

[3] The Heckman Probit Selection model uses probit estimation in the select stage (and not logistic estimation) because of the need to assume that the error term of the selection equation is normally distributed (Wooldridge 2003).

Multivariate Analysis

that are not known a priori, we need to disentangle these effects as well as the effects of persons C, D, and so on in the household. It is a violation of the fallacy of endogeneity not to separate the effects of the predictor and outcome variables.

We use three techniques to overcome this problem. First, we lag B's influence, by measuring it at a time prior to the outcome variable, at time, $T - 1$, when A's partisan support and preference are assessed at time, T.[4] As more than two persons compose many households, we extend the effect beyond B, by including the partisan preferences of all others in the household as well. Second, we use an instrumental variable in place of a direct measure of B's (and the others') partisanship. Here, we need to show for example that even if B's (and the others') partisanship may not be independent of A's party choice, a composite measure of other specified characteristics may be a reasonable substitute. This is justified by the claim that there is no reason to expect these traits to respond to A's partisanship. It is, therefore, a useful "instrument." Each of these solutions works; each has limitations, and we use each and all of them in this volume.

Each solution has different consequences for the analysis. Lagging B's (and the others') partisanship is simple and useful, but it relies on an artificial distinction between points in time. The state of knowledge of statistics permits instrumental variables to be used only in models, which address one dependent variable at a time, and none of these models addresses partisan constancy. Because each model has a possible flaw (as well as obvious strengths), we use several models during the course of the book. In this chapter, we show the results of those statistical models, which assess the effect of others in the household by lagging their partisanship. Here, we are establishing the plausibility of our most basic claim. In later chapters, where we detail these reciprocal relationships, we use instrumental variables in two-stage probit and three-stage linear probability models.

In order to assess household partisanship, we designate one person as the respondent and categorize all the others who live there as part of the household context.[5] Here, we show that the partisan preferences of those other persons in the household affect the head's partisan choices. In Chapter 4, we replace this distinction – useful because it is our first cut into the analysis – with a more nuanced analysis of household partners, incorporating explicitly the simultaneity of reciprocal interpersonal influence

---

4 This follows the lead offered by studies of "peer effects" (see for example Manski 1993; Sacerdote 2000).
5 Each survey lists persons in the household by number, and in each case we choose number 1 as the respondent. These are almost always the head of household, or the oldest male; 0.42 of the German sample and 0.48 of the Britons are heads of households.

within the household into our statistical models. In Chapters 5 and 6, we examine partisan interactions among parents and children. In order to describe the partisan composition of the region or state, we aggregate the partisan preferences of all persons in the sample who live there.

MODELING PARTISAN SUPPORT AND PREFERENCE IN GERMANY

We begin with the Germans. First, we list hypotheses taken from traditional and still useful studies of partisan support – whether or not a person inclines towards a party – in established democracies. Hypotheses about *life cycle effects* expect the youngest cohorts to display the lowest levels of party support. In Germany, generational effects would also expect the oldest persons to retain traces of their socialization under the Nazis and display relatively low levels of partisanship (see for example Norpoth 1984). We distinguish three age cohorts: 30 and younger 31–50, and 50 and older. Two other variables measure *political time*: the number of years from the start of the survey, so as to assess the impact of the secular decline in partisan support, and a dummy variable for each year in which there is a national election taps the tendency for the rate of partisan preference and support to rise at elections. We also include variables that assess the effects of *economic position*, household income and its mean-centered second-order effect (household income squared, both controlling for the number of persons present); *education*, well known to influence partisan support; and *broad social contexts* (membership in unions, Catholic religious identification – measured dichotomously; frequency of religious attendance, taking part in voluntary organizations, and socializing). Finally, we include a measure of the respondent's general level of *political interest* as a predictor of the decision to support a political party.[6]

Other variables allow us to tap our interest in partisan support in the family and neighborhood. One measures the level of *aggregate partisan support in the respondent's region*. Another factor – of primary theoretical importance – depicts the level of *support among other members in the respondent's household*. As noted, we recognize the potential problem of endogeneity – mutual influence within a household – by lagging by one year the predictor variables that detail the number of other persons in the household who also name one of the parties.

---

[6] This is important on grounds of theory (the well-known expectation that political interest influences partisan support) and method. In order for the Heckman Probit Selection model to be identified, one variable must be included in the selection equation that is *not* also found in the outcome equation (Sartori 2003; Wooldridge 2003). Political interest meets both demands.

Multivariate Analysis

Our measure partisan support within households varies along a scale based on the number of persons present who support a party. We except the relationship between this and the other predictor variables and partisan support to be positive and statistically significant, except for age cohort and household income squared, for each of which we except to observe a curvilinear pattern. Furthermore, our theory expects to find that the level of partisan support in the household has an especially strong influence on the probability of supporting a political party.

The second set of analyses focuses on the choice of party – the SPD or the CDU/CSU. Here, we draw on classic theories of party preference as well as variables that tap our interest in the effect of household partisanship on the party selected. Following the debates in the literature on the social bases of German partisanship, we apply again the measures of *social class and religion*; these describe a person's generalized position in the social and economic structure; in addition to the variables mentioned previously, we also include whether or not the respondent is a Catholic (expected to choose the CDU/CSU). Here, the *partisanship of the region* measures the percentage that prefers each party and *household partisanship* describes the net support for one or the other of the major parties among the others in the family. In addition, the party's national strength, election year, and distance from the start of the survey assess a host of political factors that affect the rise and fall of a party's popularity. Note as well that *membership in a union* is not solely a measure of social class; it also highlights the probability that a person will encounter other persons of the same social class and – by implication – partisan preference. Various *worries about the economy* may be associated with partisan choices, where concerns move persons away from the governing party (the CDU/CSU in most years of the survey). These provide a set of variables for the analysis of partisan choice.

Tables 3.1 and 3.2 present the results of the Heckman Probit Selection models. The coefficient, which measures strength of association, is in the first column, the standard error is next, and the Z-score, which assesses the level of statistical significance of each predictor variable, is in the third.

We begin our discussion of the results by noting that the presence of others in the household who incline towards a party always increases the probability of party support. Further, persons who live with others, most of whom name a particular political party, have a very strong probability of also preferring that party. Household partisanship strongly affects party support and preference, net of the other predictor variables.

Similarly, party support and preference respond to the distribution of these factors in the respondent's region of residence, if not as strongly as they reflect the household's politics. Note as well that the absence or presence of Catholic religious identification, the frequency of church attendance, and union membership – measures of social ties and social

Partisan Families

Table 3.1. *Partisan Choice for SPD – Heckman Probit Selection Model*

PANEL A: SPD CHOICE

	Coefficient	S.E.	Z-Score	Significance
Probit Model with Sample Selection				
Family SPD choice (lag 1 year, left to right)	−0.95	.05	−17.8	.000
SPD choice in region, percentage	2.19	.51	4.3	.000
SPD choice in nation, percentage	0.52	.91	0.6	.573
Political interest	0.05	.07	0.7	.517
Union member	0.49	.06	7.7	.000
Catholic	−0.27	.06	−4.3	.000
Religious attendance	−0.08	.02	−3.4	.001
Education (ref: no degree)				
Secondary school degree	−0.01	.24	−0.1	.962
Nonclassical degree	−0.29	.24	−1.2	.238
Technical degree	−0.28	.26	−1.1	.276
Academic high school	−0.48	.25	−1.9	.053
Other degree	−0.24	.34	−0.7	.490
Goldthorpe measures (ref: not working)				
High service class	−0.11	.10	−1.1	.273
Low service class	0.06	.08	0.7	.471
Routine nonmanual	−0.09	.12	−0.7	.482
Simple office worker	−0.03	.10	−0.3	.761
Self-employed, 1–20 co-workers	−0.60	.13	−4.5	.000
Self-employed, no co-workers	−0.21	.18	−1.2	.252
Skilled manual	0.11	.09	1.2	.226
Unskilled manual	0.09	.09	1.1	.285
Farm worker	0.46	.29	1.6	.118
Self-employed farmer	−1.57	.38	−4.1	.000
Household income	0.00	.00	−3.2	.001
Household income squared	0.00	.00	2.3	.020
Worried about economy	0.13	.03	4.3	.000
Worried about personal finances	0.10	.03	2.9	.004
Age (ref: 16–30)				
Middle aged (31–50)	0.04	.07	0.5	.618
Older than 50	0.01	.09	0.1	.900
Female	0.03	.07	0.5	.629
Years from start of panel	0.01	.01	2.5	.012
Year 1987	−0.03	.04	−1.0	.328
Year 1990	0.03	.04	0.7	.471
Year 1994	−0.04	.04	−1.0	.301
Year 1998	−0.01	.04	−0.2	.882
Constant	−1.34	.50	−2.7	.008

Multivariate Analysis

PANEL B: SUPPORT FOR ANY PARTY

	Coefficient	S.E.	Z-Score	Significance
Selection Model				
Family support for a party (lag 1 year)	0.47	.03	16.5	.000
Party support in region, percentage	2.21	.27	8.1	.000
Political interest	0.50	.03	20.1	.000
Union member	0.15	.04	3.5	.001
Volunteer activities	0.05	.02	2.8	.004
Social activities	0.05	.02	3.0	.003
Catholic	0.02	.04	0.6	.578
Religious attendance	0.02	.01	1.5	.144
Education (ref: no degree)				
Secondary school degree	−0.01	.13	−0.1	.948
Nonclassical degree	−0.06	.13	−0.5	.628
Technical degree	−0.03	.15	−0.2	.862
Academic high school	−0.02	.13	−0.2	.856
Household income	0.00	.00	−1.8	.068
Household income squared	0.00	.00	2.2	.303
Age (ref: 16–30)				
Middle aged (31–50)	0.11	.05	2.2	.025
Older than 50	0.26	.06	4.7	.000
Years from start of panel	0.00	.00	0.6	.583
Year 1987	0.03	.03	1.2	.215
Year 1990	0.01	.03	0.2	.810
Year 1994	0.00	.03	0.2	.866
Year 1998	0.04	.03	1.2	.248
Constant	−2.70	.23	−11.9	.000

$n = 25,758$
Log pseudolikelihood $= -22,583.7$
Statistical probability of rho statistic $= 0.048$
Observations clustered on person identification code

class – also influence partisanship in every model. Only these variables consistently and strongly influence the two elements of partisanship, and they do so after controlling for the effects of the other variables in each of the models, all of which derive from long-standing theories of German partisanship. The rho statistic is positive and statistically significant, indicating a positive association between partisan support and choice in Germany (see the Appendix for more on this measure).

Consider the variables that influence partisan support, if not preference. Here, political interest takes pride of place. The expected measures of social context – membership in unions, rate of religious attendance, activity in voluntary organizations, as well as differences in age cohort – positively influence the probability of support for a party. Note too that

Partisan Families

Table 3.2. *Partisan Choice for CDU/CSU – Heckman Probit Selection Model*

PANEL A: CDU/CSU CHOICE

	Coefficient	S.E.	Z-Score	Significance
Probit Model with Sample Selection				
Family CDU choice (lag 1 year left to right)	0.99	.05	19.8	.000
CDU choice in region, percentage	2.02	.83	2.4	.015
CDU choice in nation, percentage	2.60	.91	2.8	.004
Political interest	0.12	.06	2.1	.040
Union member	−0.45	.07	−6.1	.000
Catholic	0.44	.07	6.8	.000
Religious attendance	0.10	.02	4.6	.000
Education (ref: no degree)				
Secondary school degree	0.34	.25	1.4	.164
Nonclassical degree	0.52	.25	2.1	.040
Technical degree	0.46	.27	1.7	.086
Academic high school	0.28	.25	1.1	.275
Other degree	0.66	.36	1.8	.066
Goldthorpe measures (ref: not working)				
High service class	0.15	.11	1.4	.168
Low service class	−0.06	.08	−0.7	.497
Routine nonmanual	0.10	.13	0.8	.438
Simple office worker	0.01	.11	0.1	.910
Self-employed, 1–20 co-workers	0.48	.13	3.8	.000
Self-employed, no co-workers	0.04	.18	0.2	.818
Skilled manual	0.00	.09	0.0	.965
Unskilled manual	−0.15	.10	−1.5	.144
Farm worker	−0.60	.35	−1.7	.086
Self-employed farmer	0.57	.39	1.5	.145
Household income	0.00	.00	1.2	.223
Household income squared	0.00	.00	−0.9	.348
Worried about economy	−0.09	.03	−2.9	.004
Worried about personal finances	−0.12	.04	−3.2	.001
Age (ref: 16–30)				
Middle aged (31–50)	0.09	.08	1.2	.222
Older than 50	0.33	.09	3.7	.000
Female	−0.14	.07	−2.0	.047
Years from start of panel	0.01	.00	1.4	.174
Year 1987	0.03	.04	0.7	.478
Year 1990	−0.05	.04	−1.3	.211
Year 1994	0.07	.04	1.8	.078
Year 1998	0.03	.04	0.8	.416
Constant	−2.41	.38	−6.4	.000

Multivariate Analysis

PANEL B: SUPPORT FOR ANY PARTY

	Coefficient	S.E.	Z-Score	Significance
Selection Model				
Family support for a party (lag 1 year)	0.48	.03	16.8	.000
Party support in region, percentage	2.20	.28	7.9	.000
Political interest	0.50	.02	20.1	.000
Union member	0.16	.04	3.5	.000
Volunteer activities	0.04	.02	2.4	.019
Social activities	0.05	.02	2.8	.005
Catholic	0.02	.04	0.5	.589
Religious attendance	0.02	.01	1.6	.114
Education (ref: no degree)				
Secondary school degree	0.01	.13	0.1	.957
Nonclassical degree	−0.04	.13	−0.3	.733
Technical degree	−0.01	.15	0.0	.971
Academic high school	−0.01	.13	−0.1	.940
Household Income	0.00	.00	−1.7	.086
Household income squared	0.00	.00	1.9	.058
Age (ref: 16–30)				
Middle aged (31–50)	0.11	.05	2.2	.030
Older than 50	0.26	.06	4.6	.000
Years from start of panel	0.00	.00	0.6	.570
Year 1987	0.03	.03	1.1	.281
Year 1990	0.00	.03	−0.1	.952
Year 1994	0.01	.03	0.3	.730
Year 1998	0.04	.03	1.2	.237
Constant	−2.71	.23	−11.8	.000

$n = 25,602$
Log pseudolikelihood $= -22,038.6$
Statistical probability of rho statistic $= 0.038$
Observations clustered on person identification code

household income, but not the respondent's level of education or occupation, matter. These results do not reaffirm the claims that a secular decline in the level of partisanship and/or the ability of national elections to stem this process (see Figure 2.1) influence partisan support in any one year. Measures of social class and social interaction and a person's level of political interest join with the political effects of social intimates and others in the region on this dimension of partisanship.

Examining the "outcome" portion of Table 3.1 (the party named) reaffirms the importance of many of these variables. Here, whether or not the respondent is a Catholic joins with religious attendance and union membership to influence partisan preference, net of all the other variables. Age cohort, election year, and the measure of secular decline in

Partisan Families

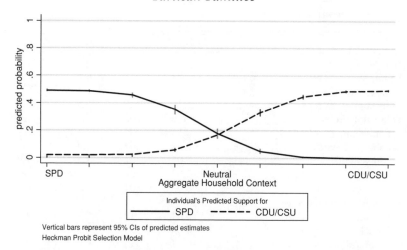

Figure 3.1. Household Influence on Partisan Choice in Germany

partisanship have no influence on this outcome. Variations in education and household income matter, as do differences in occupation. Note as well that economic worries distinguish support for the two dominant parties (Sanders and Brynin 1999, but see Anderson, Mendes, and Tverdova 2004; Erikson 2004; Evans and Anderson 2001; Johnston et al. 2005, who suggest that this is an endogenous relationship). All these variables pale in importance when compared to the effects of partisan preferences of others in the household and region. Partisanship responds to general social contexts, but it especially shows the effects of variations in the partisan preference in intimate social units.

In the concluding step of our analysis, we use postestimation techniques, drawing on the results of the Heckman Probit Selection models in order to illustrate the substantive implications of the results.[7] Here, we present the predicted probabilities of selecting one or another of the major parties, initially holding all the other predictor variables in the table at their means. We then calculate the predicted probabilities of selecting one or another of the major parties along the range of possible values of the predictor variable: Partisan context of the household.

Figure 3.1 highlights the powerful impact of a person's immediate political context on decisions to support one or the other of the major parties. Consider first the mean predicted probability of support for each party: SPD, 0.26 and the CDU/CSU, 0.23. Both panels show the "S"-shaped nature of the relationship between the distribution of partisan preference

[7] We describe the technique for developing postestimation probabilities from a Heckman Probit Selection model in the Appendix.

Multivariate Analysis

in the household and the head's choice. Almost all of the impact occurs between a net of two partners in favor of one or the other of the parties. The relationship is not linear, and so continuing to accumulate those who support (or deny) a party in the household does not much change the relationship beyond the effect of two persons. Note too the strength of this relationship. Having a net balance of two in the household in favor of the party doubles the predicted rate beyond the mean. Note too that the negative relationship is even stronger. Having a net negative balance of one drops the predicted probability of choosing the party below half of the mean. A net negative balance of two drops the predicted probability of supporting the party to close to zero. The partisan choices of heads of households respond strongly to the preferences of the persons who live with them.

MODELING PARTISAN SUPPORT AND PREFERENCE IN BRITAIN

The determinants of partisan support and choice in Britain do not much differ from those in Germany. Again, we examine the responses of heads of households who are ever interviewed. We also apply a host of variables that measure age cohort, political time, social and economic contexts (including a measure of subjective social class), economic concerns, and political interest, as well as several political variables, year from the start of the survey, election year, and national party strength. We also include regional party support and strength. And again, we apply a lagged measure of household partisanship. This scale assigns negative scores for net preference for Labour and positives for net preference for the Tories. We use much the same predictor variables in the British analyses as we do for the German, and we apply the same theoretical expectations.

The reader will not be surprised to find that here too household partisan preferences and the distribution of partisanship in the respondent's region have powerful effects on whether or not he or she picks any party and on the party selected. This holds, even while controlling for factors which have long been known to influence partisan support in Britain – various measures of social class and religion.

Note too the complexity that emerges in Tables 3.3 and 3.4. With few exceptions, each of the predictor variables influences the dependent variables, net of the effect of each of the others. While the preferences of family members and others in the region strongly influence partisan choice, and while partisan support among these persons and the level of the head of household's political interest strongly affect partisan support, the various measures of religion, religiosity, union and other organizational membership, social class identification, education, occupation, and income almost always influence both the selection variable – party support – and

Partisan Families

Table 3.3. *Partisan Choice for Labour – Heckman Probit Selection Model*

PANEL A: LABOUR PARTY CHOICE

	Coefficient	S.E.	Z-Score	Significance
Probit Model with Sample Selection				
Family Labour choice (lag 1 year, left to right)	−0.76	.03	−24.4	.000
Labour choice in region, percentage	2.21	.18	12.0	.000
Labour choice in nation, percentage	−2.03	.30	−6.7	.000
Political interest	0.19	.03	7.4	.000
Union member	0.28	.05	6.1	.000
Subjective social class	−0.23	.02	−12.7	.000
Church of England	−0.20	.04	−5.4	.000
Religious attendance	−0.05	.01	−4.0	.000
Education (ref: no schooling)				
University degree	−0.20	.08	−2.6	.010
Upper school qualification	−0.39	.06	−7.0	.000
Lower school qualification	−0.31	.05	−6.2	.000
Goldthorpe measures (ref: not working)				
High service class	−0.21	.05	−4.3	.000
Low service class	−0.18	.06	−3.1	.002
Petty bourgeoisie	−0.47	.07	−6.3	.000
Skilled manual worker	−0.11	.06	−1.8	.070
Unskilled manual worker	−0.02	.06	−0.3	.751
Household income	0.27	.19	1.4	.154
Household income squared	−0.08	.08	−1.1	.292
Finances worse than last year	0.00	.02	0.0	.984
Age (ref: 16–30)				
Middle age (31–50)	−0.14	.05	−2.9	.004
Older than 50	−0.17	.06	−2.6	.009
Female	−0.04	.04	−0.9	.354
Years from start of panel	−0.01	.00	−2.7	.007
Year 1992	0.05	.03	1.8	.077
Year 1997	0.01	.02	0.5	.648
Year 2001	0.07	.03	2.7	.007
Constant	−0.49	.19	−2.6	.009

PANEL B: SUPPORT FOR ANY PARTY

	Coefficient	S.E.	Z-Score	Significance
Selection Model				
Family support for a party (lag 1 year)	0.25	.02	13.8	.000
Party's support in region, percentage	1.64	.31	5.3	.000

Multivariate Analysis

	Coefficient	S.E.	Z-Score	Significance
Political interest	0.42	.01	29.1	.000
Union member	0.12	.03	3.8	.000
Organizational memberships	0.04	.01	3.5	.000
Church of England	0.12	.03	4.7	.000
Religious attendance	−0.02	.01	−1.9	.054
Education (ref: no schooling)				
University degree	0.17	.06	2.8	.006
Upper school qualification	0.14	.04	4.0	.000
Lower school qualification	0.04	.03	1.1	.290
Household income	−0.12	.14	−0.9	.377
Household income squared	0.01	.05	0.1	.899
Age (ref: 16–30)				
Middle aged (31–50)	0.06	.03	1.8	.075
Older than 50	0.41	.04	13.7	.000
Female	−0.04	.03	−1.6	.110
Years from start of panel	0.03	.01	5.5	.000
Year 1992	−0.01	.03	−0.2	.824
Year 1997	−0.07	.03	−2.7	.007
Year 2001	−0.04	.03	−1.6	.100
Constant	−0.31	.13	−2.4	.017

$n = 39,430$
Log pseudolikelihood = −35,848.5
Statistical probability of rho statistic = 0.00
Observations clustered on person identification code

the outcome variable – party choice. As in Germany, age influences the probability of naming a party and the party chosen. Older respondents are more likely to pick the Tories than Labour. Here too there is no evidence of a secular decline in partisanship; indeed, we find a positive relationship between the year of the survey and the probability of naming a party, net of the effect of the other variables. The rho statistic at the bottom of each table indicates that there is a correlation between the processes by which respondents support a party and the particular party preferred for Labour, but not the Conservatives.[8] Even as the relationship between partisanship and social class, religion, and age reaffirm well-known and long-standing arguments, we underline that measures of immediate social contexts have especially powerful effects on the probability of naming a party and the particular choice.

The postestimation analysis reaffirms the importance of household effects (Figure 3.2). Here, the mean predicted probability of support for

---

[8] As we found in Germany, however, there are no important substantive differences between the parameter estimates of the Heckman Probit Selection models and simple probit models applied to each variable.

Partisan Families

Table 3.4. *Partisan Choice for Conservatives – Heckman Probit Selection Model*

PANEL A: CONSERVATIVE PARTY CHOICE

	Coefficient	S.E.	Z-Score	Significance
Probit Model with Sample Selection				
Family Conservative choice (lag 1 year, left to right)	0.91	.04	25.5	.000
Conservative choice in region, percentage	1.76	.31	5.8	.000
Conservative choice in nation, percentage	0.89	.71	1.3	.209
Political interest	0.02	.05	0.4	.681
Union member	−0.34	.06	−5.9	.000
Subjective social class	0.22	.02	11.2	.000
Church of England	0.32	.04	7.9	.000
Religious attendance	0.03	.01	2.0	.049
Education (ref: no schooling)				
University degree	−0.15	.09	−1.6	.102
Upper school qualification	0.29	.06	4.8	.000
Lower school qualification	0.23	.06	4.1	.000
Goldthorpe measures (ref: not working)				
High service class	0.26	.06	4.7	.000
Low service class	0.17	.07	2.6	.010
Petty bourgeoisie	0.44	.08	5.6	.000
Skilled manual worker	0.11	.07	1.4	.153
Unskilled manual worker	0.06	.07	0.8	.441
Household income	−0.18	.22	−0.8	.400
Household income squared	0.00	.09	−0.1	.964
Finances worse than last year	−0.08	.03	−2.7	.007
Age (ref: 16–30)				
Middle age (31–50)	0.22	.06	3.9	.000
Older than 50	0.38	.08	4.7	.000
Female	−0.04	.04	−0.8	.411
Years from start of panel	0.03	.01	2.5	.011
Year 1992	0.01	.05	0.1	.890
Year 1997	0.00	.03	0.1	.927
Year 2001	−0.10	.03	−3.3	.001
Constant	−1.75	.37	−4.7	.000

PANEL B: SUPPORT FOR ANY PARTY

	Coefficient	S.E.	Z-Score	Significance
Selection Model				
Family support for a party (lag 1 year)	0.25	.02	13.3	.000
Party's support in region	1.69	.33	5.2	.000

Multivariate Analysis

	Coefficient	S.E.	Z-Score	Significance
Political interest	0.42	.01	28.9	.000
Union member	0.11	.03	3.6	.000
Organizational memberships	0.03	.01	2.8	.005
Church of England	13.00	.03	5.0	.000
Religious attendance	−0.02	.01	−1.7	.094
Education (ref: no schooling)				
University degree	0.17	.06	2.8	.005
Upper school qualification	0.14	.04	4.0	.000
Lower school qualification	0.04	.03	1.2	.234
Household income	−0.12	.14	−0.9	.381
Household income squared	0.01	.05	0.1	.910
Age (ref: 16–30)				
Middle aged (31–50)	0.06	.03	1.9	.052
Older than 50	0.42	.04	10.9	.000
Female	−0.04	.03	−1.5	.136
Years from start of panel	0.03	.01	5.5	.000
Year 1992	0.00	.03	−0.2	.863
Year 1997	−0.08	.03	−2.7	.007
Year 2001	−0.05	.03	−1.8	.066
Constant	−0.30	.14	−2.2	.026

$n = 39,515$
Log pseudolikelihood = −34,209.3
Statistical probability of rho statistic = 0.85
Observations clustered on person identification code

Labour is 0.32 and 0.19 for the Conservatives. As we see in Germany, there is also an "S"-shaped relationship between household partisanship and partisan preference in Britain. Again, the negative effect is more powerful than the positive one. It takes only a net of two persons in the household who favor one of the major parties to make it just about certain the head does not name that party. A net positive effect of two for Labour doubles the respondent's predicted probability of naming Labour, when compared to the mean level of predicted Labour preference, whereas it triples it for the Conservatives. Again, as is present in the German data, this pattern holds for both parties, no matter the relative strength of Labour vis-à-vis the Tories.

These results condition how we proceed in our subsequent models of partisanship and electoral behavior. Simply put, efforts to account for partisan support and preference in Germany and Britain should include measures of the partisanship of household members and others in the region and more general and abstract indicators of social class and religion, the classic structural bases of politics in these democracies. Furthermore, measures of political interest need to be included as well. Obviously,

Partisan Families

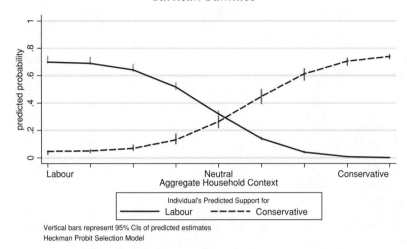

Figure 3.2. Household Influence on Partisan Choice in Britain

they find a place in the analysis of partisan support, but where we model only partisan preference, we add this variable because of its indirect effect on choice, via party support. We also include the measures of age cohorts. All subsequent models include these variables.

PARTISAN CONSTANCY IN GERMANY AND BRITAIN

Recall that partisan constancy joins two conceptually related phenomena: partisan choice at one point in time and the frequency of that preference over time. The first – partisan choice – is a simple yes or no response: the respondent either does or does not name Party A/B. Aggregating, or "counting," the number of times that the party is named during the available opportunities over time measures partisan constancy (providing a count score for each respondent; see Figures 2.7 and 2.8 for a description of the variable in Germany and Britain). Using these scores counsels that the analysis moves beyond OLS regression analysis to use a count model.[9] Because the count results presented in Figures 2.7 and 2.8 indicate that there are an excess of zeros and because the data are overdispersed (i.e., the variance of the dependent variable in each case is greater than the

---

9 Standard OLS (ordinary least squares) regression analysis requires that the dependent variable conceivably take any real value on the number line from negative to positive infinity, a requirement that our dependent variable clearly violates, given that it can only take integer values from zero to the maximum possible – seventeen in the German case, and eleven in Britain (corresponding to the number of years of the panel survey in each country).

Multivariate Analysis

mean), we employ the Zero-Inflated Negative Binomial (ZINB) model (Winkelmann 2000).

The ZINB model produces two sets of parameter estimates: (1) the "inflated" (or binary) portion that provides information about the predictors of the likelihood of remaining at zero (i.e., never supporting the party – corresponding to partisan choice) and (2) the "count" portion that determines the number of times that a respondent chooses the party, given a positive response on partisan choice. This portion models partisan constancy. Assessments of the relative impact of each of the predictor variables on partisan choice and constancy accompany each table. In addition, using each model's parameter estimates, we employ Stata/SE8.2's *prvalue* command to predict the number of times a household head would support a particular political party, given the chosen values of the predictor variables. For ease of interpretation across the two cases, we transform the raw number of the count into a percentage by multiplying it by 100, providing the portion of the opportunities that a person names the party. The ZINB model enables us to specify the absolute and relative ability of critical variables to predict the rate of party preference. In this portion of our analysis, we examine persons who are in all waves of GSOEP and BHPS, so as to model partisan constancy.

Again, we model the partisan behavior of heads of household.[10] Here, not only do we lag the partisan characteristics of others in the household, but we do the same for all the other predictor variables as well. In Germany, these come from 1985 and 1986, and in Britain, we take them from 1991 and 1992.[11] We use the responses in all the remaining years in order to measure partisan choice and constancy. The analysis examines the extent to which a set of a person's characteristics at time T structures his or her subsequent political choices. Given our theoretical stance, we pay particular attention to the net distribution of partisan preferences in the household. We also include the most important variables from our previous analyse: measures of the party's strength in the respondent's region of residence, social class (union membership in both Germany and Britain and subjective social class in Britain), religion (Catholic

---

10 In GSOEP, heads amount to 0.46 of the respondents, and in BHPS, they account for 0.52 of all persons who respond to the partisanship questions in all waves. Not surprisingly, these persons are much more likely to be older than the sample means (by about four years in each survey) and are also more likely to be men. With regard to partisan preferences, however, they differ from the full sample only by displaying slightly higher rates of naming a party in any one year and over time as well as a higher rate of partisan constancy between any two adjacent years.

11 As a check, we ran more complex models, which contain all the relevant variables from the Heckman Probit Selection analyses. Here, we present only the variables with the strongest impact on partisan constancy.

Table 3.5. *Partisan Choice and Constancy in Germany (1985–2001) – Zero-Inflated Negative Binomial Regression Model*

	SPD						CDU/CSU					
	Inflate Model			Count Model			Inflate Model			Count Model		
	Coefficient	Z-Score	Significance	Coefficient	Z-Score	Significance	Coefficient	Z-Score	Significance	Coefficient	Z-Score	Significance
Family SPD choice 1985/86, left to right	0.82	13.2	.000	−0.15	−7.4	.000						
Family CDU/CSU choice 1985/86, left to right							−0.77	−12.82	.000	0.2	8.3	.000
SPD choice in region 1985, percentage	−3.08	−2.2	.027	0.61	1.1	.286						
CDU choice in region 1985, percentage							0.81	0.43	.665	−0.75	−0.8	.401
Political interest 1985/1986	0.03	0.6	.565	0.06	2.8	.005	−0.14	−3	.003	0.07	2.7	.008
Union member 1985	−0.82	−5.4	.000	0.16	2.9	.004	0.42	2.91	.004	−0.11	−1.5	.139
Catholic 1990	0.21	1.5	.141	−0.04	−0.6	.541	−0.39	−2.78	.005	0.17	2.5	.013
Religious attendance 1990	0.24	3.6	.000	0.00	0.0	.990	−0.25	−3.67	.000	0.05	1.7	.083
Age in 1985 (ref: 16–30)												
Middle aged (31–50)	−0.15	−0.8	.428	0.02	0.2	.815	−0.06	−0.34	.773	0.13	1.4	.167
Older than 50	−0.02	−0.1	.938	0.00	0.0	.986	−0.07	−0.35	.723	0.26	2.5	.011
Constant	0.43	0.7	.459	3.38	13.7	.000	0.94	1.54	.122	3.26	10.7	.000

$n = 1{,}276$
Log likelihood $= -4{,}159.9$

$n = 1{,}276$
Log likelihood $= -3{,}846$

Multivariate Analysis

or not in Germany and Church of England or not in Britain, and the frequency of church attendance in both), age cohort, and level of political interest. Note that because the model looks for the effect of each predictor taken in the first or second year of the survey on partisan preference and constancy, this is a particularly demanding test for the hypotheses. The results sustain the absolute and relative explanatory importance of the partisan preferences of family members.

Both models tell the same simple story (see Tables 3.5 and 3.6): the partisan choice and constancy of the head of the household are directly and strongly affected by the partisan preferences of others in the household. Only this variable influences both the binary and count processes for each of the four parties. Political interest always influences partisan constancy. In other words, high levels always generate higher predicted counts – greater partisan constancy – for each of the parties, independent of other effects. Union membership in both populations affects partisan choice for both parties and also constancy for the SPD and Labour, but not the other parties. The effects of religion and the rate of church attendance mirror this pattern: positive for both dependent variables for the CDU/CSU and the Conservatives and negative for the SPD and Labour. Note that the partisan characteristics of the respondent's region influence the choice of the party, but not the rate by which the party is named over time. Also, the more interested is the head of household in politics at T, the higher is the rate of partisan constancy, net the effect of the other variables in both models.

Consider as well the ways that variation in age affects partisan choice and constancy. Persons in the youngest cohort are the least likely to name a particular party at a given point in time. These results are consistent not only with long-standing findings, which show age differentials with regard to partisanship, but also with the results of our Heckman Probit Selection models. At the same time, however, the youngest set of respondents is no less constant than the others with regard to partisan choices. In Germany, the *count* portions of the ZINB models show no differences among the age categories; in Britain, they show slight differences between the oldest (but not the middle category) and youngest respondents with regard to partisan consistency. We will return to this topic in Chapter 5.

There are multiple ways to display the relative power of each of the predictor variables on partisan constancy. The Stata/SE8.2 command, *prvalue* can be used to simulate and calculate predicted (expected) counts of the dependent variables based on different values of the predictor variables and the parameter estimates obtained from the ZINB analyses (not displayed, but available from the authors). These simulations demonstrate the extraordinary impact of household partisanship on the head's partisan preference over time. In the simulations, the head of a large family all of

Table 3.6. *Partisan Choice and Constancy in Britain (1991–2001) – Zero-Inflated Negative Binomial Regression Model*

	Labour						Conservatives					
	Inflate Model			Count Model			Inflate Model			Count Model		
	Coefficient	Z-Score	Significance	Coefficient	Z-Score	Significance	Coefficient	Z-Score	Significance	Coefficient	Z-Score	Significance
Family Labour choice 1991/92, left to right	0.65	16.5	.000	−0.10	−8.7	.000						
Family Conservative choice 1991/92, left to right							−0.82	−18.6	.000	0.09	6.0	.000
Labour choice in region 1991, percentage	−3.29	−6.6	.000	0.70	4.2	.000						
Tory choice in region 1991, percentage							−3.16	−5.2	.000	0.16	0.6	.546
Political interest 1991/1992	−0.09	−3.0	.003	0.07	6.9	.000	−0.09	−3.1	.002	0.08	5.9	.000
Union member 1991	−0.34	−2.9	.004	0.08	2.2	.029	0.27	2.2	.029	−0.07	−1.4	.151
Subjective social class 1991	0.29	5.9	.000	0.29	−3.9	.000	−0.29	−5.6	.000	0.04	1.9	.063
Church of England 1991	0.23	2.4	.015	0.00	−0.1	.934	−0.46	−4.6	.000	0.06	1.6	.122
Religious attendance 1991	0.15	3.2	.001	0.00	0.1	.946	−0.08	−1.6	.105	−0.01	−0.6	.544
Age in 1991 (ref: 16–30)												
Middle aged (31–50)	0.01	0.1	.955	−0.01	−0.2	.816	−0.21	−1.5	.134	0.04	0.6	.562
Older than 50	−0.27	−2.0	.049	0.10	2.2	.031	−0.14	−1.0	.332	0.2	3.1	.002
Constant	0.89	3.8	.000	3.47	38.7	.000	2.62	8.9	.000	3.47	25.4	.000
	$n = 2{,}504$						$n = 2{,}504$					
	Log likelihood = −8,064.2						Log likelihood = −6,296.6					

Multivariate Analysis

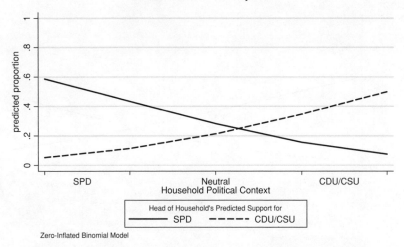

Figure 3.3. Household Influence on Partisan Constancy in Germany

whom support Party A is almost certain to do so again and again. None of the other variables comes close to this level of influence. Even as other measures of broad social context affect partisan choice or constancy, what matters most for partisan constancy is what other members of the family prefer. No matter the differences that separate the two established democracies, these patterns hold.

A graph of the relationship between the different values of household partisan preference and predicted rates of partisan constancy highlights the importance of this variable. In order to ease the interpretation, we display the effects of five ordered categories of household preference on partisan choice. These vary from −2, a net of two persons who favor the SPD or Labour, to +2, a net of two persons in favor of the CDU/CSU or the Conservatives; we keep all the other predictor variables at their means.

First, note the predicted mean values for the heads of households: SPD, 0.28; CDU/CSU, 0.21; Labour, 0.32; and the Conservatives, 0.21.[12] In other words, based on the parameter estimates of our model, and setting all the predictor variables (including household partisanship) at their mean levels, heads of household in Germany are predicted, on average, to choose the SPD 0.28 of the time (which is equivalent to 3.92 times over the seventeen years), and so on for the others. Figures 3.3 and 3.4 illustrate the results for each country. In both Germany and Britain, persons who live with two others who support Party A in the first two years

[12] Recall that we transform the raw counts into rates, in order to ease the comparison between Germany and Britain.

Partisan Families

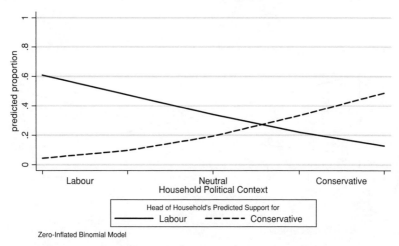

Figure 3.4. Household Influence on Partisan Constancy in Britain

of the surveys are twice as likely as the mean head of household to support the party over time. Conversely, living with two others who did not name the party in the first two years implies never supporting the party. Furthermore, this analysis implies the characteristics of persons certain to always choose a party: high supportive partisanship in the household, maximum levels of political interest, and maximum levels of one of the measures of social class. There are relatively few such persons in each sample. These patterns derive from and reinforce the bounded nature of partisanship. Most people never name one of the dominant parties, and they vary the rate of naming its rival. We return to this point many times in our analysis.

Our analysis highlights variables that significantly raise or lower the probability of partisan support, preference, and constancy in Germany and Britain. Again and again, the models show that variance in the partisan decisions responds directly to variation in the political preferences of others in the household. At the extremes of political cohesion in family (and presumably other social) networks, political choices are easy to predict: these respondents are twice as likely as the mean respondent to support Party A/B, and they never pick Party B/A.

Note the imbalance between the social logic that affects partisan choice of one party and the rejection of the other. The presence of only one net preference for Party A/B ensures that Party B/A will not be chosen. This asymmetry in the relationship is of fundamental importance: it makes it easier to eliminate one party from the decision set than to select a party from that group. It provides the social basis for bounded partisanship.

Multivariate Analysis

THE WAY FORWARD

Repeatedly finding that members of a household strongly influence the partisanship of the household's head certainly supports our argument. At the same time, it raises warning flags. Perhaps the intimate ties that connect husbands and wives and parents and children and others who live together are so strong that the household, and not the individual person, is the appropriate unit of analysis (Becker 1964; 1981; 1985 offers the classic statement). Alternatively, perhaps the household effect masks another even more fundamental source that precedes the formation of the household. After all, the principle of associative mating implies that persons who resemble each other marry (or live with) each other. Is there not reason to suppose that the political preferences precede the household bond? Perhaps the emphasis on social connections is an illusion; persons who live together are almost certain to share the same location in the social structure, and so perhaps the effects that we perceive are just the result of individuals in the same social locations deriving the same political consequences from these shared positions.

Our answers to these questions point the way forward. In Chapter 1, we elaborate a set of theoretical claims that disentangles the persons in the household and details how they relate to each other. These give us theoretical grounds for arguing that influence within the household is neither determined nor illusory. In the next chapters, we highlight the interactions between household partners (usually husbands and wives) and among parents and children, showing that households do not subsume their members and demonstrating precisely how persons influence each other. In these chapters, we move past the claim that there is a household effect to show how it works. As we do this, we substantiate the importance of social interactions, not just shared social locations.

Before we move ahead, we prefigure the empirical portion of the analysis by examining the partisan contexts of the heads of household, the focus of this chapter. Put most simply, relatively few of them live in families with a distinctive partisan cast. Consider a summary measure of the net number of persons who support each of the major parties, during the years of each survey (the samples used in the Heckman Probit Selection models). In both Germany and Britain, nearly 0.65 of the heads live in households scored at zero (no one else who lives there supports a party or there is a balance between the Social Democrats and the Christian Democrats/Socials, or Labour and the Conservatives, or the respondent lives alone). Nearly 0.19 (Germany) or 0.20 (Britain) of the sample reside among persons who favor the SPD/Labour. Of the German heads of households, 0.17 live among others with a net favorable score for the CDU/CSU, and in

Partisan Families

Britain the rate does not reach 0.15 for the Tories. German and British households are complex political units.

These data are congruent with others that describe political variations in the immediate social networks of respondents (Huckfeldt, Johnson, and Sprague 2004; Schmitt-Beck 2003; Stoker and Jennings 2005; Zuckerman and Kotler-Berkowitz 1998). Most households are not cohesive political units, and therefore, they should not be considered to be a single unit of analysis. In turn, accounting for the variation in the political cohesion of households becomes the next stage of our analysis. Finally and of significance too, husbands and wives assert their independence from their partners' influence; they deny that they simply reflect the political views of their mates (Beck and Jennings 1991 and Kenny 1994 on the United States and Miller, Wilford, and Donaghue 1999 on Northern Ireland).

Finally, our analysis does not imply the presence of rigid and widespread political cleavages. It does imply that Germany and Britain are characterized by the presence of persons who vary in the extent to which they live surrounded by persons who share the same political likes and dislike, thoughts and emotions. These variations influence partisan support, choice, and movement.

4

Bounded Partisanship in Intimate Social Units: Husbands, Wives, and Domestic Partners

> Therefore, a man shall leave his mother and father and cling to his wife, and they shall be one flesh.
> Genesis 2:2

> It is not good for man to be alone.
> Genesis 2:18

> God always pairs off like with like.
> *The Iliad*, XVII, 1:218

> Others, on the contrary, hold that all similar individuals are mutually opposed.
> Aristotle, *Nicomechean Ethics* 1155a: 214–16

Families stand as the primary social units; households are recurrent locations for social interactions. No matter that there are other partners in conversation – parents, children, friends, neighbors, workmates, and acquaintances – people talk to their spouses and domestic partners. In all these intimate social units, not only words but groans, smiles, frozen silences, grimaces, laughter, and shouts convey preferences. No matter the level of interest of politics in the household, the greater the number of thoughts exchanged, the more they discuss politics (Walsh 2004).[1]

---

[1] Studies of conversations are the standard source for explorations of the effects of immediate social contexts on political preferences (see Conover, Searing, and Crewe 2002; Huckfeldt and Sprague 1995; Pattie and Johnston 1999; 2000; 2001; Walsh 2004; Zuckerman and Kotler-Berkowitz 1998; Zuckerman, Kotler-Berkowitz, and Swaine 1998; and Zuckerman, Valentino, and Zuckerman 1994). The literature on marital homophily specifies the networks-based approach (Dogan 1967; Glaser 1959–60; Katz and Lazarsfeld 1955; March 1953–4, and Merton 1957 are classic sources; also see the discussion in Chapter 1). For a recent application to electoral choice in Britain, see De Graaf and Heath (1992), and for an early version of our analysis, see Zuckerman, Fitzgerald, and Dasović (2005).

Partisan Families

Families are the primary locus of political discussions; here, national politics enters the personal space of each citizen (Glaser 1959–60; Hays and Bean 1992; March 1953–4; Stoker and Jennings 1995; Verba, Schlozman, and Burns 2005; and Zuckerman and Kotler-Berkowitz 1998). Absent information on these issues in GSOEP or BHPS, we bring illustrative evidence from other sources. For example, in 1987, more than 0.75 of Britons report that the person with whom they most frequently discuss politics is a family member who lives in their household; 0.66 single out their spouse or live-in partner as that person; 0.30 talk politics with no one but family members, and less than 0.05 do not include a family member among the two persons with whom they most frequently discuss politics (Zuckerman, Kotler-Berkowitz, and Swaine 1998). Data from the Local Participation Study in the United Kingdom for the year 2001 indicate that persons are more likely to discuss politics with their husbands, wives, and other household partners, than their parents, children, neighbors, workmates, and social friends.[2] As persons who live together exchange ideas and opinions about politics, they state, test, reinforce, and reformulate their political preferences.

The Germans and Britons who compose the couples we study are bounded partisans. As we show in Chapter 2, most of them never name one of the major parties, and they vary their choice of the other large party – sometimes choosing it and sometimes not. They are more consistent with regard to the party that they do not name than with their preferred party. Also, very few go from party to party. Partisan choice and constancy vary according to specific social contexts – preferences within households and the more distant and abstract associations that are linked to social class and religion. Political interest affects these choices indirectly through its impact on party support. These findings provide the baseline for the analysis of the dynamics of partisanship between household partners.

In this chapter, we extend the analytical thread of our argument by demonstrating that spouses and domestic partners influence each other's partisan preference; the relationship is reciprocal, if not symmetrical, and similar, if not precisely the same in the two countries. Furthermore, it is not influenced by the asymmetrical distribution of political interest between partners. Contrary to what some would suppose, the male partner – usually the more interested in politics – does not dominate his partner's

---

[2] Colin Rallings and Michael Thrasher direct the study that was administered in 2002, UKDA #4849. Here are the mean levels of discussion partners, where 0 is "none" and 9 is "every day:" spouses and partners (4.6); parents (1.9); children (2.2); neighbors (2.0); workmates (2.5); and social friends (4.0). We would like to thank the United Kingdom Data Archive, at the University of Essex, for providing access to these data; they bear no responsibility for our analysis.

partisan preferences and choices. Rather, spouses who are bounded partisans maintain each other. In the next two chapters, we expand this to include children as well as mothers/wives and fathers/husbands, as we further detail the pattern of partisan influence within German and British households. In the conclusion, we suggest that these reciprocal relationships rest on another reciprocal relationship that links political parties and households.

In the first part of this chapter, we show that each partner affects the other's partisan preferences, net of any other factor. They influence each other's partisanship, no matter each person's own propensity to name a political party that derives from social class, religion, or any other generalized social location or identity. In the second section, political similarity between husbands and wives emerges as general, but not point specific. Household partners usually stay together on the same side of the political divide, but because each moves over time between Party A/B and no party, they are as likely as not to support the same party at any moment. They are, however, much more likely to agree on the party that they do *not* name. Households, as such, are cohesive with regard to the party not named, not the party chosen.

As a result, we examine the determinants of political similarity; we do not assume its presence. We ask: does political agreement between household partners derive from shared secondary characteristics, such as social class or religious identifications? Does it follow from the general principle of marital homophily, "like to like," in which persons with like political minds marry each other? Our analysis will present no evidence to support this last claim. Instead, we show that concurrence in partisan preference among partners rests on shared social characteristics, high levels of political interest, and living together for many years.

Our study of couples reports the analysis of seventeen waves in Germany (1985–2001) and eleven years of annual choices in Britain (1991–2001). Counting the responses of each set of partners who answer the survey at least once produces approximately fifty thousand couple-years in Germany (with the precise numbers varying by question). More than ten thousand couples answer BHPS's questions at least once (again with variation across the different items).

WHO AFFECTS WHOM? HUSBANDS AND WIVES AFFECT EACH OTHER'S PARTISAN PREFERENCES

Let us first address a central piece of our analysis. In Chapter 3, we show that the net partisan preferences of others in the immediate family influence both the partisan choice and constancy of heads of households,

Partisan Families

no matter the importance of other predictor variables. As we discuss there, modeling partisanship in households needs to resolve problems of endogeneity – A influences B, but B also affects A. In Chapter 3, we use a lagged predictor variable, examining the partisanship of others in the household in the year before the head responds. These analyses offer no more than a partial solution to the problem because they do not allow us to model directly the simultaneity of the mutual influence of household partners. Now, we apply instrumental variable probit models in order to sort the mutual influence of husbands and wives (or men and women who are domestic partners) on each other's partisan preferences. Our analysis specifies the strength of the effect on each strand of the dyads, which link husbands and wives, net of the influence of other explanatory variables.

The models are simple and straightforward. Partisan choice, the selection of a particular party, we suggest, is a function of particular characteristics of the person and the partisan preference of the spouse. Drawing on the results of the models in the previous chapter, we again include measures of social class and religion. In Germany, where there is no measure of subjective class identification, we use union membership and specific categories of occupations as a surrogate, and in Britain we use subjective social class as well as union membership and occupation.[3] In Germany, we measure religion by whether or not the respondent is a Catholic. In Britain, we distinguish between those who are Church of England and those who are not; these are the measures of religious identification with the most direct links to partisanship. In both cases, we also include the frequency of attendance at religious services. As is well known and as we reaffirm in the previous chapter, variables that indicate the working class and the absence of religion link persons to the SPD or Labour; measures of middle class and religiosity (particularly Catholic in Germany and Anglican in Britain) are associated with a preference for the CDU/CSU or the Conservatives. These are important – if obvious – predictors of partisan choice. Our analyses also include the measure of political interest because of its demonstrated impact on the probability of partisan support and, indirectly therefore, on partisan choice. In order to separate the effects of the husband's and wife's partisanship, we employ an instrumental variable – a measure that correlates with spouse A's partisan preference but not with B's choice or any of the predictor's of partner B's preference. For Germany, we employ union membership and various occupations

---

[3] The measures of Goldthorpe occupations are derived from logit analyses of party choice that include the variables with statistically significant results in the "select" portion of the Heckman Probit models in Chapter 3. In addition, we also tested other more elaborate models, with no substantive benefit to the analysis.

Bounded Partisanship – Spouses and Domestic Partners

Table 4.1. *Reciprocal Effects on SPD Choice – Instrumental Variable Probit Model*

	Coefficient	S.E.	Z-Score	Significance
Panel A: Husband's SPD Choice				
Wife's SPD choice	1.74	.23	7.7	.000
Husband's political interest	0.18	.02	11.5	.000
Husband's union membership	0.42	.04	11.4	.000
Husband is Catholic	−0.24	.03	−8.5	.000
Husband's religious attendance	−0.06	.01	−5.1	.000
Husband's occupation				
High service class	−0.11	.03	−3.6	.000
Self-employed 1–20 co-workers	0.01	.04	0.3	.732
Self-employed no co-workers	−0.57	.06	−10.1	.000
Skilled manual	0.09	.02	3.6	.000
Unskilled manual	0.10	.03	3.2	.001
Constant	−1.41	.06	−22.6	.000

Endogenous variable: wife's SPD choice
Instrumental variables: wife's union membership, wife's high service class, wife's self-employed 1–20 co-workers, wife's self-employed no co-workers
$n = 30,002$

	Coefficient	S.E.	Z-Score	Significance
Panel B: Wife's SPD Choice				
Husband's SPD choice	2.39	.10	24.7	.000
Wife's political interest	0.35	.01	24.2	.000
Wife's union membership	0.29	.03	8.5	.000
Wife is Catholic	−0.26	.02	−12.2	.000
Wife's religious attendance	−0.09	.01	−8.8	.000
Wife's occupation				
High service class	−0.20	.08	−2.5	.013
Self-employed 1–20 co-workers	0.03	.03	1.0	.299
Self-employed no co-workers	−0.17	.08	−2.3	.024
Constant	−2.09	.05	−42.5	.000

Endogenous variable: husband's SPD choice
Instrumental variables: husband's union membership, husband's high service class, husband's self-employed no co-workers, husband's skilled manual, husband's unskilled manual
$n = 30,064$

(but neither of the religious measures), and in Britain, union and subjective social class join to form the instrument (but again not the measures of religion). In both cases, the dependent variable is dichotomous, measuring the choice of party. This produces one model for each partner for each party in each country, eight in total. Again, we use Stata/SE 8.2.

We begin with the GSOEP data. There are two panels in each of Tables 4.1 and 4.2. Table 4.1 examines the choice of the SPD, first for the

Table 4.2. *Reciprocal Effects on CDU/CSU Choice – Instrumental Variable Probit Model*

	Coefficient	S.E.	Z-Score	Significance
Panel C: Husband's CDU/CSU Choice				
Wife's CDU/CSU choice	2.25	.45	5.0	.000
Husband's political interest	0.17	.03	6.9	.000
Husband's union membership	−0.27	.04	−6.8	.000
Husband is Catholic	0.32	.05	6.9	.000
Husband's religious attendance	0.05	.02	1.9	.056
Husband's occupation				
High service class	0.09	.03	2.6	.011
Self-employed 1–20 co-workers	0.08	.05	1.6	.106
Self-employed no co-workers	0.33	.06	5.8	.000
Skilled manual	0.01	.04	0.2	.865
Unskilled manual	−0.10	.05	−1.9	.056
Constant	−1.91	.04	−45.4	.000

Endogenous variable: wife's CDU/CSU choice
Instrumental variables: wife's union membership, wife's high service class, wife's self-employed 1–20 co-workers, wife's self-employed no co-workers
$n = 29{,}759$

	Coefficient	S.E.	Z-Score	Significance
Panel D: Wife's CDU/CSU Choice				
Husband's CDU/CSU choice	2.88	.15	19.5	.000
Wife's political interest	0.22	.02	13.2	.000
Wife's union membership	−0.28	.05	−6.2	.000
Wife is Catholic	0.08	.03	3.0	.003
Wife's religious attendance	0.11	.01	10.1	.000
Wife's occupation				
High service class	0.05	.08	0.6	.556
Self-employed 1–20 co-workers	−0.09	.04	−2.6	.009
Self-employed no co-workers	0.07	.07	1.0	.344
Constant	−2.47	.04	−57.9	.000

Endogenous variable: husband's SPD choice
Instrumental variables: husband's union membership, husband's high service class, husband's self-employed no co-workers, husband's skilled manual, husband's unskilled manual
$n = 29{,}817$

husband and then the wife, and Table 4.2 models the selection of the CDU/CSU, again beginning with the husband. Each displays strong support for the claim that domestic partners influence each other's partisan choices, after controlling for predictor variables linked to occupation, union membership, religion, and political interest. Note, however, that the husband has a much stronger effect on the wife's partisan choice than does the female partner on the male, for both the SPD and the CDU.

Bounded Partisanship – Spouses and Domestic Partners

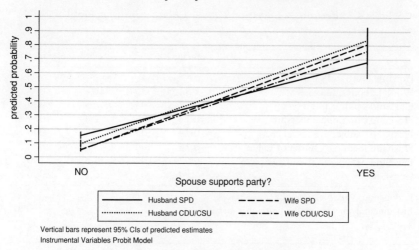

Figure 4.1. Reciprocal Effects of Partisan Choice in Germany

In Figure 4.1, we display the impact of husbands and wives on each other's partisan selection for each party based on the postestimation probabilities.[4] Here, we hold each of the other predictor variables at their mean values, and we locate two cases: where the respondent's partner names the party and where he or she does not. No matter the relative importance of social class for the husband and religion for the wife, they influence each other. Comparing the slope lines (as well as the z scores in Tables 4.1 and 4.2) indicates that the male partner has more effect than the wife. Furthermore, the husband's partisan choice has more influence on the wife than her own social class and religious characteristics, but her impact on him is not as strong as whether or not he is a Catholic, belongs to a union, and his level of political interest.

Tables 4.3 and 4.4 display the instrumental probit models for the two major British parties. Here, the wife has greater relative influence than in Germany. In turn, her partisan preference outweighs any of the measures of his social class or religion and political interest. Figure 4.2 transforms these results into a more easily interpreted figure, by focusing on the reciprocal effects of household partners, after holding the other predictor variables at their means. In Britain, wives have a greater impact on their male partners than do the husbands on their spouses.

Persons who live together influence each other's partisan choices, net of the other determinants of partisanship. More generally, we view our

---

4 In the Appendix, we describe how we construct the postestimation probabilities, where Stata/SE8.2's *prvalue* command does not apply.

Table 4.3. *Reciprocal Effects on Labour Choice – Instrumental Variable Probit Model*

	Coefficient	S.E.	Z-Score	Significance
Panel A: Husband's Labour Choice				
Wife's Labour choice	2.82	.30	9.5	.000
Husband's political interest	0.22	.03	6.9	.000
Husband's union membership	0.13	.05	2.4	.019
Husband's subjective social class	−0.23	.03	−8.3	.000
Husband member Church of England	−0.16	.04	−3.6	.000
Husband's religious attendance	0.02	.02	1.3	.190
Constant	−1.94	.07	−26.8	.000

Endogenous variable: wife's Labour choice
Instrumental variable: wife's union membership
$n = 7,775$

	Coefficient	S.E.	Z-Score	Significance
Panel B: Wife's Labour Choice				
Husband's Labour choice	1.91	.24	8.0	.000
Wife's political interest	0.28	.02	11.3	.000
Wife's union membership	0.30	.06	4.9	.000
Wife's subjective social class	−0.10	.03	−3.3	.001
Wife member Church of England	−0.04	.04	−1.0	.315
Wife's religious attendance	−0.06	.01	−3.8	.000
Constant	−2.02	.09	−22.3	.000

Endogenous variable: husband's Labour choice
Instrumental variable: husband's union membership
$n = 7,775$

results as an important and additional step towards validating the general claim that political learning – the exchange of political preferences – occurs within households. At the same time, mutual influence does not mean that each person's political preferences are simply a function of their intimate social partner. As is evident in the models, political interest, religion, and social class also influence partisanship, net the effects of the partner's choice. Household partners influence but are not indistinguishable from each other.

In the next portion of this chapter, we elaborate this picture of husbands and wives who resemble each other, but who do not move in lockstep. At any point in time, not even half the couples name the same political party. Instead, the reciprocal influence of German and British husbands and wives (or male and female domestic partners) emerges as each stands near, but apart from the other. They almost never support one of the major parties and then either jointly refuse to support any party or split

Bounded Partisanship – Spouses and Domestic Partners

Table 4.4. *Reciprocal Effects on Conservative Choice – Instrumental Variable Probit Model*

	Coefficient	S.E.	Z-Score	Significance
Panel C: Husband's Conservative Choice				
Wife's Conservative choice	1.84	.27	6.8	.000
Husband's political interest	0.03	.02	1.3	.194
Husband's union membership	−0.14	.05	−2.7	.008
Husband's subjective social class	0.18	.03	6.8	.000
Husband member Church of England	0.27	.05	5.9	.000
Husband's religious attendance	−0.02	.01	−1.3	.190
Constant	−1.19	.09	−13.3	.000

Endogenous variable: wife's Conservative choice
Instrumental variable: wife's union membership
$n = 7,775$

	Coefficient	S.E.	Z-Score	Significance
Panel D: Wife's Tory Choice				
Husband's Conservative choice	1.62	.24	6.8	.000
Wife's political interest	0.14	.02	5.9	.000
Wife's union membership	−0.26	.07	−4.0	.000
Wife's subjective social class	0.17	.02	7.2	.000
Wife member Church of England	0.29	.04	6.9	.000
Wife's religious attendance	0.01	.01	0.7	.505
Constant	−1.63	.07	−22.4	.000

Endogenous variable: husband's Conservative choice
Instrumental variable: husband's union membership
$n = 7,775$

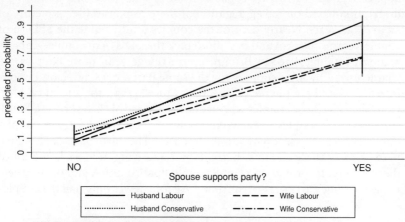

Vertical bars represent 95% CIs of predicted estimates
Instrumental Variables Probit Model

Figure 4.2. Reciprocal Effects of Partisan Choice in Britain

Partisan Families

the choices between the other major party and no party. While couples are not indivisible units, they choose singular paths; almost all household pairs move along routes that no other set of partners follows.

Later in the chapter, we also show that their political similarity derives from their shared position with regard to measures of social class and religion as well as their interactions over time. The same factors that influence partisan choice for each person also affect the rate at which couples share partisanship: (1) variations in the level of political interest – a personal characteristic, which affects the probability of supporting a party – and (2) generalized social contexts such as social class and religious locations. Political preferences do not form a component of marital homophily – "like to like," and "birds of a feather flock together" (to recall the passages from Aristotle cited in chapter one and highlighted at the opening of this chapter) – the tendency for persons to marry persons like themselves. Instead, the longer that they have been together, the more likely they are to share partisanship at a particular point in time. Combining these variables allows us to predict with statistical precision when people who live together will support the same political party and when they will not. Partisanship responds strongly but not exclusively or overwhelmingly to people's intimate social relations.

COUPLES SOMETIMES PREFER THE SAME POLITICAL PARTY

What is the precise level of agreement on the choice of a party over time? As we do in Chapter 2, we answer this question twice: among couples who are interviewed at least once and among couples whose responses appear in all the waves. We begin with the first set in Germany, offering a description based on the aggregation of these responses (i.e., a picture of an average year). Nearly 0.19 of the couples jointly name the SPD (n of couple-years $= 38,459$) and 0.17 both pick the CDU (n of couple-years $= 38,097$), totaling 0.36. In Britain in an average year, more couples are characterized by joint support for the same party: 0.18 agree on Labour ($n = 8,273$) and 0.24 agree on the Tories ($n = 8,273$), for a total of 0.42. Consider now those couples who are in every wave of each survey. GSOEP surveys 1,033 couples over seventeen waves, providing 14,462 couple-opportunities to assess partisan preferences. The partners pick the same party 0.42 of these chances. BHPS taps 1,428 couples in each and all of the eleven years, providing 15,708 couple-opportunities, and the partners name the same party 0.43 of the time.[5] All told, fewer than half the couples

---

[5] In both countries, examining whether both choose "no party" increases the rate of similarity, to 0.67 in Germany and 0.60 in Britain. Because explanations of picking "no party" differ from those for the selection of one of the major parties, we confine

Bounded Partisanship – Spouses and Domestic Partners

(about four out of ten) agree on the same major party, when they have an opportunity to express their views.

When household partners do not name the same major party, they also do not name Party A/B and its major rival Party B/A. Rather, differences in households are characterized at one point in time, by one partner naming Party A/B and the other preferring no party. In Germany, one spouse prefers the SPD and the other the CDU/CSU 0.03 of the time. In Britain, the rate is 0.04. In 0.10 of the pairs in each country, one member picks both of the major parties, at different times over the many waves of the surveys. Hardly anyone who ever chooses one of the major parties lives with someone who ever names the other.[6]

"Sometimes" is also the appropriate answer to the question, "Do couples claim the same class or religious identity?" GSOEP asks about religious but not social class identification. There, 0.41 name the same religion. In Britain, 0.57 share either a working class or middle class identification, and 0.30 name the same religion. Let us compare the rate of mixed households on each of these dimensions: 0.09 of the German households have partners with different religions (and another 0.10 have one partner with a religion and the other without). Whereas less than 0.01 of British households have partners with different religious identities, 0.34 are characterized by one claiming to identify with a religion and the other not. In turn, 0.18 report different class identities, and the same portion characterizes households in which only one member claims class identification. Couples vary in the extent to which they share religious and class identification as well as partisanship.

Consider now the partisan choices of household partners over time. Figures 4.3 and 4.4 focus on the selection of each of the major parties, summarizing the choices made by the German and British couples who respond to the partisan preference question in every wave of each survey. Couples jointly avoid one of the dominant parties and vary their joint support for the other. Note that among those partners who ever pick a party, they are equally likely to come back to it an inconsistent number of

---

this analysis to partisan choice. Note as well that Huckfeldt, Johnson, and Sprague (2005) report that 0.60 of the dyads in the 2000 American National Election Study hold the same political preference, and Petersson, Westholm, and Blomberg (1989: 312–5) present similar patterns within blocs of political parties in Sweden.

6 On a scale that varies across three locations – Party A, no party, Party B – almost all the movement occurs between the middle category and one of the two extremes. As a result, correlations are not appropriate measures of association; they always overstate the level of similarity on partisan choice. In the next chapter, we include comparisons with children in the households. There, we find more variable levels of these correlations, because many more of the youngsters support no party in any one year.

Partisan Families

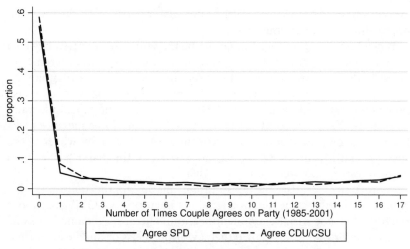

Figure 4.3. Joint Partisan Constancy in Germany

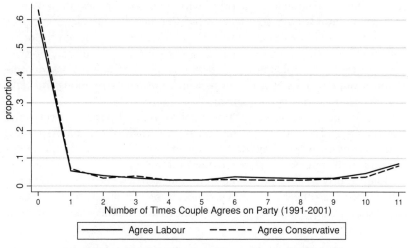

Figure 4.4. Joint Partisan Constancy in Britain

times – that is, one, two, three, four, or any number of years in the surveys. The British data show that the responses for the Tories and Labour are remarkably similar; it is barely possible to perceive the patterns of two distinct lines.

Over time, couples also travel along unique political paths. Here we apply a columnar report, in order to observe the trail or path of choices that each person in each couple makes during the years of the panel

Bounded Partisanship – Spouses and Domestic Partners

surveys. How many trails do we observe? In Germany, there are 1.2 couples per path, and in Britain, 1.3 couples per path. In both cases, eliminating those sets in which both partners always offer the same responses (almost always uniform support for one of the parties) reduces the number of paths to barely more than one per couple (1.06 in Germany and 1.08 in Britain). As is true for individuals, no couple moves on a track that mimics the relative strength of the parties over time aggregated to the level of each nation. Just as important and notwithstanding the large-scale issues that characterize national politics, each pair responds to them in ways that distinguish it from all the others. In this regard at least, each couple is its own political universe.

Persons who live with others (that is to say, most people) are bounded partisans who influence each other. This relationship usually reinforces each person's tendency to stay on one side of the national political divide. As a result, they almost never split their partisan preferences between the two major parties. Instead, each moves between naming Party A/B and claiming no partisan preference. As a result, partisan concurrence within households is a variable, and so the analysis seeks to account for the extent to which both partners make the same partisan choice and the frequency of that selection over time.

MODELING PARTISAN AGREEMENT BETWEEN HOUSEHOLD PARTNERS

Partisan concurrence between spouses, we show, is a joint function of several variables. Shared identification with or location in one of the social classes and religions offers one set of predictors. Where husbands and wives name the same religion, where they both frequently attend religious services, and where they both identify with the same social class and both are union members, they are especially likely to share the same partisanship. Living in regions in which their neighbors provide strong support for a party also increases their probability of shared preference. Note as well that the length of time that they live together directly influences the level of partisan similarity within households, net of the other predictor variables. Put simply, newlyweds are less likely to prefer the same party than are veteran couples. Because partisan agreement rests on the propensity to name a party and because political interest influences the individual's propensity to claim partisan support, we expect the variation in the level of political interest that characterizes the partners to influence the probability that they share partisan choice and consistency. Elections, secular trends of how they drift away from the parties, and the relative political successes and failures of the parties provide a general context for these microprocesses.

Partisan Families

Table 4.5. *Partisan Agreement between Husband and Wife for the SPD – Time-Series Logit Model*

	Coefficient	S.E.	Z-Score	Significance
SPD choice in region, percentage	3.55	.40	8.9	.000
SPD choice in nation, percentage	−1.09	1.32	−0.8	.409
Total political interest	0.20	.01	14.9	.000
Agree union membership	0.19	.08	2.5	.013
Agree Catholic	−0.42	.08	−5.3	.000
Agree no religious attendance	0.12	.04	3.0	.002
Years married	0.01	.00	4.6	.000
Constant	−3.41	.15	−22.7	.000

$n = 29,100$
Wald Chi^2 = 587, p = .000

Here, we assess the relationship between partisan agreement in the household and the predictor variables with a logit model that takes into account the autocorrelation that affects responses in time-series panel data. We derive a series of measures appropriate to each political party and country to test our hypotheses. In both countries, we use shared union membership, religious identification (in Germany, Catholic or not; in Britain, Church of England or not), and high religious attendance as well as partisan strength in the couple's region, and in Britain we also include shared social class identification. In order to tap political interest, which we expect to influence partisan support, and indirectly partisan choice, we measure the total level of political interest in the household. We also include the number of years that the couple has been married, a predictor that we expect to affect shared partisan support and shared partisan choice. Again, we begin with the German data.[7] Table 4.5 displays the model for agreement for the Social Democrats, and Table 4.6, for the Christian Democrats/Socials. In both models, the results are simple, clear, and reinforcing. First, each of the measures of a couple's social context has a strong effect on the probability that they will both choose the SPD or the CDU/CSU, net of all the other variables. Shared support for the party derives not simply from being married or living in the same household. Indeed, the longer that they live together, the greater is the probability of

---

[7] In order to include a measure of the number of years that the couple is married, we exclude from our analysis those persons whose marital status changes during the survey years (about 0.10 of the sample). This skews the analysis a bit: it may overlook young women in the labor force, who may be especially prone to remain single or to leave marriages (Iversen and Rosenbluth 2003). In these models, we apply Stata/SE 8.2's *xtlogit* command, in order to take account of the times-series nature of the data.

Bounded Partisanship – Spouses and Domestic Partners

Table 4.6. *Partisan Agreement between Husband and Wife for the CDU/CSU – Time-Series Logit Model*

	Coefficient	S.E.	Z-Score	Significance
CDU/CSU choice in region, percentage	7.14	.30	24.2	.000
CDU/CSU choice in nation, percentage	0.44	.50	0.9	.380
Total political interest	0.18	.01	13.0	.000
Agree union membership	−0.15	.10	−1.5	.129
Agree Catholic	0.58	.07	7.8	.000
Agree high religious attendance	0.14	.05	2.7	.006
Years married	0.02	.00	12.1	.000
Constant	−4.88	.18	−27.6	.000

$n = 26,960$
Wald $Chi^2 = 882, p = .000$

shared partisan choice (see also Beyer and Whitehurst 1976 and Stoker and Jennings 2005.) This agreement is also enhanced by shared social locations, with regard to religion and union membership, as well as the partisan distribution of others in their region of residence. Furthermore, high levels of political interest affect shared partisanship, by increasing the rate by which each partner names a party.

Two critical themes emerge from these models. These findings affirm Huckfeldt, Johnson, and Sprague's (2005) claim that political agreement in dyads is autoregressive, depending on the extent to which others in a person's social network share the particular political views. Absent these social reinforcements, the probability of agreement drops. As important, the results show that it is not the case that cohabitation by itself results in partisan agreement. Because newlyweds have relatively low rates of shared partisan choice, it is also unlikely that partisan preferences are central to the initial choice of partner.

Figures 4.5 and 4.6 displays the relationship between the probability of shared choice of each partner and the number of years that the couple has been married. With regard to agreement on each of the major political parties, it takes many years of marriage – twenty-two – in order for the couple's probability of agreement to exceed the mean. Slowly but surely, couples who live together come to resemble each other with regard to partisan choice.

BHPS permits us to offer more complex models of partisan agreement in households.[8] The British data include measures of subjective social

---

[8] In order to include a measure of the length of a marriage, we examine those who are in the survey in 1992 and/or 2001, the only two years BHPS asks the respondents how long they have been married. Those who enter after the first year and

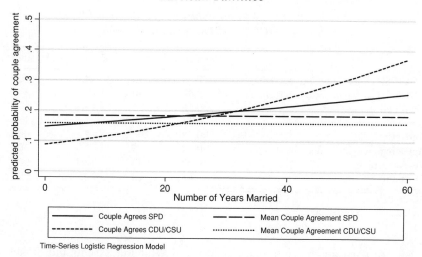

Figure 4.5. Postestimation Probabilites: Years of Marriage and Partisan Agreement in Germany

class (working class or middle class) and religious identification (we focus on membership in the Church of England). Table 4.7 displays the results for Labour and Table 4.8 for the Conservative Party. Again, high levels of political interest in the household sustain partisan agreement. Furthermore, the probability of jointly naming Labour rises with joint identification with the working class, residence in a Labour stronghold, and the party's strength in national politics. Conversely, the probability that spouses agree to support the Tories rises when both identify with the middle class and the Church of England and also live in a region of Conservative strength. Again, the length of the marriage affects the probability of jointly supporting each of the major parties. Partisan agreement in the household rests on shared social class, religious identification, and many years of marriage.

Consider now the relationship between the number of years that the couple has been living together and the probability of jointly naming

leave before the last are likely to be relatively young and mobile, more likely to support Labour. That may be why this sample shows a lower rate of shared support for Labour than in the general sample. BHPS also includes variables that assess the quality of the relationship between the two partners. Some assess decision making within the household; others measure how close each feels towards the other. These measures are difficult to interpret, and when included in a model none has a significant relationship on the probability of partisan agreement.

Bounded Partisanship – Spouses and Domestic Partners

Table 4.7. *Partisan Agreement between Husband and Wife for Labour – Time-Series Logit Model*

	Coefficient	S.E.	Z-Score	Significance
Labour choice in region, percentage	1.62	.85	1.9	.056
Labour choice in nation, percentage	3.47	1.38	2.5	.012
Total political interest	0.20	.03	5.9	.000
Agree union membership	−0.05	.22	−0.3	.801
Agree working class	0.50	.11	4.5	.000
Agree Church of England	−0.11	.13	−0.9	.383
Years married	0.01	.01	2.2	.029
Constant	−4.69	.40	−11.6	.000

$n = 5,494$
Wald Chi$^2 = 92$, $p = .000$

Table 4.8. *Partisan Agreement between Husband and Wife for Conservatives – Time-Series Logit Model*

	Coefficient	S.E.	Z-Score	Significance
Conservative choice in region, percentage	3.42	.97	3.5	.000
Conservative choice in nation, percentage	1.21	1.16	1.0	.299
Total political interest	0.11	.03	4.1	.000
Agree union membership	0.02	.19	0.1	.924
Agree middle class	0.29	.09	3.2	.002
Agree Church of England	0.32	.11	3.1	.002
Years married	0.02	.00	3.8	.000
Constant	−3.51	.26	−13.4	.000

$n = 5494$
Wald Chi$^2 = 122$, $p = .000$

Labour or the Tories. As in Germany, there is a direct positive effect, and the relationship is weaker for the party of the left. Still for both parties and in both countries, as couples stay together, they increase the probability of naming the same party. Here too the tip-over point is about twenty-two years.

SUMMARY AND IMPLICATIONS

Our analyses of GSOEP's and BHPS's survey data provide direct evidence that household members influence each other's party preferences. Each serves as a pole of attraction, keeping the partner away from one side of the national political divide. Whether or not this is the source of the

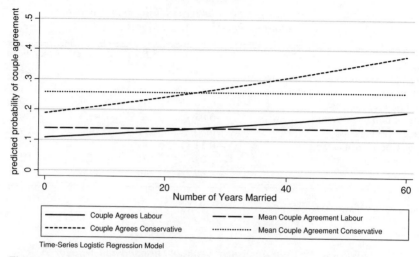

Figure 4.6. Postestimation Probabilities: Years of Marriage and Partisan Agreement in Britain

individual's proclivity not to move between the major parties, it certainly reinforces it.

Husbands affect wives, and wives affect husbands. The strength of the strands in these dyadic relationships varies across the two countries. In Germany, the male partner has a much stronger influence on his spouse, and in Britain this relationship is reversed, though the asymmetry is less marked than in Germany.

Over time, couples follow unique political paths. Hardly any two couples offer the same views of the political parties. None resembles the aggregate distribution of partisanship over time. Rather, each represents a distinct set of political choices. The political life of households is not the same as that of the nation.

Partisan concurrence within German and British households rests on shared social contexts and identities and many years of marriage. Social class and religion, long the staples of German and British social and political life, continue to matter. The more years couples live together, the more likely they are to choose the same political party recurrently. Shared party preferences do not initially bring couples together. Instead, they become more alike as they share their lives. As they live together, they offer political cues to each other and usually accept them. Absent these variables few couples name the same political party. Partisan agreement in households requires social support.

Bounded Partisanship – Spouses and Domestic Partners

Might shared partisanship rest as well on variables that are not present in GSOEP or BHPS? There is no way to provide a direct and definitive answer. There are no other surveys that allow us to examine the various elements of partisanship within households. Still, we can say that our analysis reaffirms that in households as well as in the singular decisions of persons, social class, religion, political interest, and political context influence partisan choices. In the conclusion, we present an argument that links partisanship in households to the structure of partisan politics.

There is little reason to be surprised by the finding that partners influence each other's party preferences. After all, households provide the primary locus of interaction, affection, trust, comradeship, and political discussion for their members. Additional evidence taken from BHPS underscores this general point. In four years – 1993, 1995, 1997, and 1999 – the survey asks the respondent to identify his or her closest friend: the choices include partner/spouse, various relatives, and friends. How frequently do domestic partners pick each other as their closest friends? The rate is 0.65 for the men and 0.55 for the women and, given the distribution of the sample, a total of slightly less than 0.60 of all respondents. Recall too that British couples are more likely to talk about politics with each other than with anyone else. None of this may be surprising, but it does underscore the extent to which there is reason to expect persons who live together to give and take political cues.

Not only do husbands and wives influence each other's partisan choices, but they affect each other in other ways too. Married spouses (but not household partners) influence each other in areas over which neither has much personal control, such as health (Powdthavee 2005) and mood (Huston, Caughlin, and Houts 2000). They turn to each other in areas where economic and instrumental rationality are expected to direct decisions. In Britain, business leaders frequently ask their spouses for advice. Indeed, asking one's spouse or partner is the second most frequent source of counsel (0.63); the only source more highly rated (0.69) is the managing director, chair, or chief executive (MORI Research Report 2004). Trust is the most important (0.87) reason for this choice. And note too that the health and job status of one partner influence the other's decisions about the timing of retirement (Coile 2003; Johnson and Favreault 2001; Pienta 2003). Persons in intimate social relations influence each other; they are not simply roommates, and they are not "one flesh."

In conclusion, we underline a central analytical point. Even as we focus on the importance of reciprocal relationships within households and even as we detail and model the bases of partisan concurrence in these intimate social units, it is important to remember that partisan agreement is

Partisan Families

a variable. Indeed, in most households, both partners do not share more than one or two of the critical variables: social class or religious identity, union membership, high or low levels of religious attendance, maximal levels of political interest. Most do not live in regions of maximal support for one or the other of the parties. As a result, the level of partisan agreement in households is always an empirical question; no social determinism applies.

5

Bounded Partisanship in Intimate Social Units: Parents and Children

In established democracies, young persons, like everyone else, make decisions about the political parties.[1] Neither mimicking nor ignoring their parents' partisan preferences, they, like their mothers and fathers, support or do not support a party and consider which one to name; they too make these decisions again and again. Unlike their parents, however, most young persons do not claim to be partisans. In Germany and Britain when they do name a party, they almost always pick the one selected by their mother and father. Like their parents too, young Germans and Britons are usually bounded partisans, varying their announced choices between Party A/B and no partisan preference, rarely moving between the two parties. And during the years of the surveys, as their parents – particularly their mothers – increasingly announce that they support no party, the rate by which their children deny that they support a party also rises.

The social logic of partisanship suggests a series of hypotheses to account for political influence across generations within families: (1) Frequent interactions and high levels of trust induce members of households to influence each other. (2) Where two members of the household agree, partisan influence is especially high. (3) The relative distribution of past learning in the household makes children especially likely to take partisan cues from their parents. (4) Because mothers interact more frequently with their children than do fathers and because members of the younger generation are more likely to trust their mothers than their fathers, she has more influence over the children than does her husband, no matter his higher level of political interest. (5) In turn, the father's relative advantage

---

[1] Classic statements in the literature of political socialization and partisanship include Beck and Jennings (1991); Easton and Dennis (1976); Greenstein (1965); Hess and Torney (1968); Hyman (1959); Jaros (1973); Jennings and Niemi (1968; 1981); Tedin (1974). See Sapiro (2004) for a discussion of this literature and a review of recent research. For an early application to Germany, see Baker (1974).

in political interest does not affect this process. (6) Differences in the relationship between children and each of the parents suggest that the young persons are highly likely to influence the mother, even where they have no effect on the father.

POLITICS DOES NOT DIVIDE PARENTS AND CHILDREN, EVEN ADOLESCENTS

There is nothing exceptional about the absence of partisan conflict between parents and offspring. Achen's (2002) picture of widespread clashes between generations does not emerge from research in developmental and adolescent psychology. "Relationships with parents remain the most influential of all adolescent relationships and shape most of the important decisions confronting children, even as parents' authority over mundane details (e.g., attire, hairstyle) wanes" (Collins and Laursen 2000; 2004, 337, citing Steinberg, 2001; Steinberg and Silk 2002; see also Sargent and Dalton 2001 and Bantle and Haisken-DeNew 2002 for the effects of parents on addictive behavior among teens). Music and clothing are mundane matters, akin to "galoshes and garbage," the field's label for the topics of insignificant clashes between parents and children (Steinberg 2005 and Steinberg and Silk 2002).[2] Parents and children usually get along. Indeed, BHPS's sample of adolescents reports that 0.55 are "completely happy" with their families and that 0.92 place themselves on the "happy" end of a seven-point scale. It is not surprising that on most matters, including politics, adolescents learn from their parents.

GSOEP and BHPS data underscore the similarities between parents and children regarding the substantive themes of politics, religion, and social class. Tables 5.1 and 5.2 present a series of polychoric correlations among the youngest persons in the samples (ages 16–20) and their parents. Drawing on the responses of persons who reply to the partisanship questions at least once during the surveys, the tables display the mean relationship among teens and their mothers and fathers with regard to partisan choice; level of political interest; religious identification and attendance; as well as social class identification in Britain. In each case, the teens strongly resemble their parents, especially their mothers. They may never listen to their parents' music and they may not always pay attention to their parents'

---

[2] In this context, it is useful also to note the industries that devote massive resources to influence teens' choices with regard to clothing and music (for a survey that focuses on marketing trends in Europe, see the Euromonitor International Web site: http://www.euromonitor.com/Marketing_to_children).

Bounded Partisanship – Parents and Children

Table 5.1. *Agreement among Mothers, Fathers, and Children in Germany – Polychoric Correlation Coefficients*

	Mother and Child	Father and Child	Mother and Father
SPD choice	0.63	0.53	0.79
CDU/CSU choice	0.64	0.61	0.79
No party support	0.43	0.30	0.64
Political interest	0.33	0.29	0.36
Protestant	0.95	0.81	0.75
Catholic	0.97	0.90	0.83
Religious attendance	0.61	0.62	0.71

Table 5.2. *Agreement among Mothers, Fathers, and Children in Britain – Polychoric Correlation Coefficients*

	Mother and Child	Father and Child	Mother and Father
Labour choice	0.58	0.55	0.73
Conservative choice	0.68	0.61	0.75
No party support	0.34	0.30	0.44
Political interest	0.33	0.25	0.30
Subjective social class	0.46	0.37	0.52
Church of England	0.54	0.39	0.63
Roman Catholic	0.88	0.77	0.72
Religious attendance	0.64	0.64	0.68

political views, but when they choose a political party, they almost always name the same one that their mother and father choose, and so it is too for their religious and social class identities.

Children follow their parents' partisanship, we argue, because they – like almost everyone else – usually take the political cues of trusted loved ones with whom they frequently interact. When those cues are consistent and when the young persons are interested in politics, they almost always reproduce their parents' partisanship, especially when others in the region agree. They never support Party A/B, when their parents do not name A/B, and they usually pick B/A, when mother and father do so. It would be surprising, indeed, shocking, if this were not the case. Instances of generational conflict in politics, such as those which rocked Germany and other Western nations in the late 1960s, stand out for their uniqueness. These are not the norm; they cannot be. After all, as young persons enter the electorate, they have already learned a lot from their parents, they frequently interact with them, and they usually have no a priori reason to reject their parents' cues. In the established democracies that we are studying, Germany (1985–2001) and Britain (1991–2001), the partisan preferences of the young follow that of their parents.

Our perspective directs attention to the reciprocal relationships that characterize households. There is no reason to suppose that the influence flows only from parents to offspring, especially as the children get older (see the special issue of *New Directions for Child and Adolescent Development*, Smetana 2005). Certainly, 29 year olds have different relationships with their parents than do 16 year olds. The exchange of political cues moves through dyadic relationships between each parent and child, between the parents, and among the three-way tie of mother, father, and offspring. And so, in the remainder of this chapter we expand the definition of young person to include all persons who are no more than 29 years of age.[3]

This chapter addresses a set of related questions about the partisan choices of these young Germans and Britons. What induces them to support a party? Which party? How constant over time and consistent across parties are their preferences? Where useful, we contrast some of their answers with those of their parents, but we also refer the reader back to the descriptions in Chapter 2 for additional comparisons with the full samples. We then account for the decision to support a party and the related choice of party. As in Chapter 3, the application of Heckman Probit selection models helps us to answer these questions. Our analysis displays the strong impact of parents on their children. In turn later in the chapter, we use linear probability models (three-stage least squares estimations for systems of simultaneous equations) to detail the reciprocal influences within households. Again and again, we find that mothers matter more than fathers. The end of this portion of our empirical story returns us to the volume's theoretical perspective: persons who live together influence each other's political choices. Here, we show that mothers stand at the center of households, influencing and being influenced by their husbands and children. In the conclusion, we explore the basis of this strong relationship.

Like the previous chapters, two data sets structure the analyses of each country. The first includes young persons and their mothers and fathers, who respond in any of the many waves of the two surveys. This enables us to aggregate responses in two ways: in an average year, during the period 1985–2001 in Germany and 1991–2001 in Britain and in each of the years. The second data set examines young persons who live with their

---

[3] Note that most Germans and Britons even in this expanded age category are still living at home with their parents.

Table 5.3. *Partisan Choice across Generations in German and British Households*

	Germany		
	Mothers	Fathers	Children
Chooses SPD	0.25	0.29	0.17
Chooses CDU/CSU	0.27	0.31	0.14
Supports no party	0.40	0.32	0.59
	Britain		
	Mothers	Fathers	Children
Chooses Labour	0.30	0.37	0.20
Chooses Conservatives	0.19	0.24	0.11
Supports no party	0.38	0.29	0.61

parents in the first year of each panel survey and follows them all for the remaining years.[4]

THE PARTISANSHIP OF YOUNG PERSONS AND THEIR PARENTS

How do young persons and their mothers and fathers distribute themselves among the partisan choices in an average year of these surveys? The panels in Table 5.3 present the responses for each country. The most prominent finding is that in both countries, the young are more likely to claim no partisanship than any other option. Note too that German mothers and fathers display lower levels of partisan support than do their British counterparts, and that German mothers display the lowest levels among the parents. The responses also indicate that about 0.09 of the young Germans support one or the other of the smaller parties (usually the Greens). In Britain, about 0.10 of the young distribute their support among the Liberal Democrats, the Scottish and Welsh Nationalists and other smaller parties. Children are much less likely than their parents to name one of the major parties. Indeed, it is not too strong to say that "no party" is the expected response from young persons, when asked about their partisan support.

---

[4] We remind the reader that GSOEP and BHPS provide direct information on persons 16 years of age or older in each household. Other examples of analyses of partisanship across generations that utilize separate interviews for each respondent – not recollections by children of their parents' party preferences – include Beck and Jennings (1991); Jennings and Niemi (1968; 1981); Ventura (2001); and Westholm (1991; 1999). Kroh and Selb (2005) use GSOEP data to analyze political socialization as well.

Partisan Families

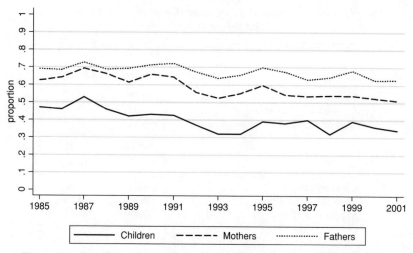

Figure 5.1. Trends in Aggregate Partisan Support in German Households

GSOEP and BHPS allow us to examine trend lines in two ways. Observing these patterns among those who answer the surveys at least once describes representative samples of Germans and Britons. The second presents the responses of those who answer partisanship questions every year. These data also allow us to examine whether or not young persons take on partisan support as they grow older and are increasingly exposed to electoral politics (see Converse 1969 for the classic statement of this hypothesis). We have two ways to examine the extent to which the youngsters come to resemble their parents.

First, we display trend lines on partisan support. Figure 5.1 shows the mean level of partisan support in each year among German children and their mothers and fathers (the number of persons who answer is always several thousand). There are persistent and slight declines over time among the young persons and their mothers in partisan support. Indeed, the lines remain nearly parallel during the years of the surveys, and the increases at election years (1987, 1990, 1994, and 1998) barely punctuate this trend. Note how similar they are. Here, we see aggregate-level evidence of the association between mothers and their offspring that receives individual-level attention in this chapter and the next. Note as well that by the end of our period, barely half the German mothers claim to support a party, well below the rate of the data's first year. Increasingly over time, young Germans are raised in households in which their mothers are as likely as not to display partisanship. We return to this point in the concluding chapter.

Bounded Partisanship – Parents and Children

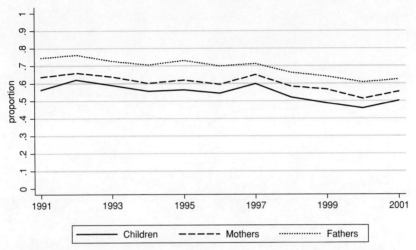

Figure 5.2. Trends in Aggregate Partisan Support in British Households

Figure 5.2 examines the same theme in Britain. Here, we find mothers moving generally halfway between their children and their spouses, and here there is evidence of secular decline in partisanship in both generations. By the end of our period, British mothers are only slightly more likely than their German counterparts to support a party. Note that the elections (1992, 1997, and 2001) are associated with increases in the portion that announce support for a political party. During the decade of the 1990s, both young persons and their mothers are increasingly likely to decline to support a political party.

Next, we examine trends in both sets of households among persons who are in all waves of each survey (see Figures 5.3 and 5.4). Again, young persons consistently display lower levels of partisan support than do parents. Here too mothers move between their husbands and their children, but among those who are in all waves, they are closer to their partners than their offspring. As important, the trend lines in each country do not come together over the years of the survey. Even as these young persons move from ages 16–32 to 33–46 in Germany and from 16–29 to 30–40 in Britain, their levels of partisanship do not much change and do not approach those of the parents. Indeed, in 2001, the levels of partisan support in the two generations hardly differ from their starting points. GSOEP and BHPS provide no evidence that mere exposure to electoral politics by itself enhances the rate of partisan support. In these established democracies, aggregate levels of partisanship do not grow during the decade or more observed. These results conform to the observation

Partisan Families

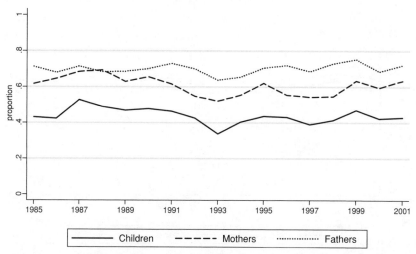

Figure 5.3. Trends in Partisan Support in German Households (among persons in all waves)

of a generalized and secular decline in European democracies (see for example Dalton and Wattenberg 2000).

THE DYNAMICS OF MICROPARTISANSHIP AMONG YOUNG PERSONS AND THEIR PARENTS

Youngsters, like their parents, are bounded partisans. Examining those who are in every wave of the surveys, we see that most never name one of the major parties and most vary their choice of the other large party – sometimes choosing it and sometimes not. They are more consistent with regard to the party, which they do not choose than with regard to their preferred party. Figures 5.5 and 5.6 present strikingly similar patterns in both countries for both generations; there is a sharp spike at the zero count and then a drop to relatively few selections at each of the other count points. Few (0.16 in Germany and 0.12 in Britain) ever pick each of the two parties, during the years of the surveys. These figures display the relative success of Labour among young persons compared to the Tories and to the two German parties. Among young persons, bounded partisanship implies a very high incidence of naming no party.

In order to detail the micropatterns, we again arrange the responses into a columnar report. In Germany, 0.05 of the youngsters who are in all waves pick the SPD or CDU/CSU every time, and another 0.08 always select no party. The others also display unique paths for each person.

Bounded Partisanship – Parents and Children

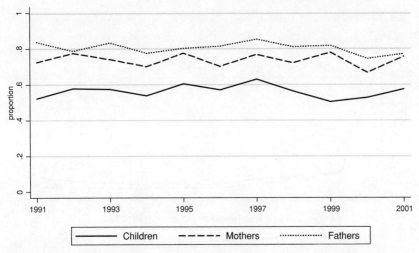

Figure 5.4. Trends in Partisan Support in British Households (among persons in all waves)

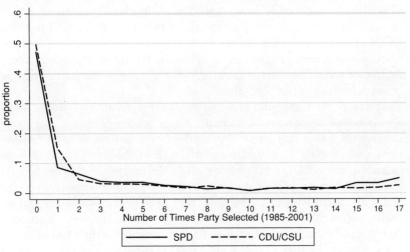

Figure 5.5. Bounded Partisan Choice among Young Persons in Germany

No one offers responses that follow the aggregate trend of the two parties (the pattern NNNNSNNNNNNNNN). Because "no party" is the plurality choice of young Germans in every wave, their answers more closely mirror national results than do older Germans or young Britons.

In Britain, 0.08 of the youth always prefer Labour, 0.05 always choose the Conservatives, and 0.15 persistently name no party. They also display

Partisan Families

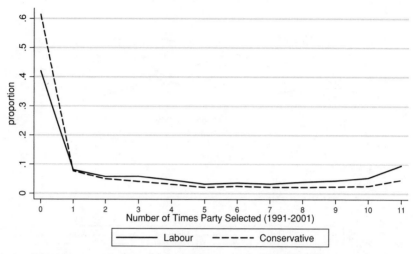

Figure 5.6. Bounded Partisan Choice among Young Persons in Britain

remarkably diverse preferences over time, on average 1.1 persons per path. Consider also the case of persons who mimic the aggregate trend, which shows a shift from a plurality for the Tories to Labour to no party in 2001: CCCLLLLLLLN. How many are there with the precise pattern? There is not one young person (and only four among the rest of the population)! Aggregate trends are the result of individual decisions, but hardly anyone behaves like the national results.

As noted, for the young persons, bounded partisanship implies a relatively high rate of naming no party. Those who ever choose the Social Democrats, name that party 0.39 of the time; they select the Christian Democrats/Socials 0.05, and no party, 0.48 of the time. Children who ever choose the CDU/CSU pick it 0.38 of the time; they choose the SPD 0.08 and no party 0.51 of the time. In Britain, young persons who ever pick Labour do so 0.54 of the time; they prefer the Tories 0.07 and no party 0.35 of the time. Those who name the Conservatives at least once do it 0.34 of the time; they cross to Labour 0.05 of the time and name no party 0.23 of the time. These youngsters are more amenable to selecting one of the other parties, particularly the Liberal Democrats. Comparing the results in Chapter 2 for the full samples, we see that young persons are no more likely to move between the major parties than are their parents. They differ by displaying higher rates of choosing no party.

As we show in the previous chapter, hardly any of the parents in a household divide between the two major political parties, but most mothers and fathers do not support the same party at a single point in time. In an average year, fewer than 0.03 of the German households are characterized

Bounded Partisanship – Parents and Children

by each parent naming a different major party; in fewer than 0.42 does each parent support the same major party; in 0.21, one parent names either the SPD or the CDU/CSU and the other names no party; and in 0.23 neither names a party. In Britain in an average year, less than 0.04 of the households are characterized by each parent supporting a different major party; in 0.41 both name the same major party (usually Labour); in 0.29 one chooses no party and the other a major party, and in 0.15 both deny having a party preference. German and British children come of political age and continue to live in households that vary in political cohesion.

During the years of these panel surveys, young Britons and Germans do not rebel against the partisan preferences of their parents. The GSOEP and BHPS data offer hardly any evidence of generational conflict. On average, 0.05 of the German youngsters name the SPD, when both of their parents support the CDU/CSU or when one parent names the CDU and the other names no party. More striking, less than 0.03 of the German youngsters name the Christian Democrats/Socials when both of their parents support the SPD or where one names the Social Democrats and the other selects no party. The British results are not much different. Where both parents prefer the Tories, 0.05 of the youngsters name Labour; in families in which both parents name Labour, none of the children (less than 0.01) prefers the Conservatives. In both countries, young persons almost never choose a major party that their parents do not prefer (and see Ventura 2001 and Westholm 1999 for parallel evidence from Israel and Sweden).

CHILDREN, PARENTS, AND PARTISANSHIP

In this section, we account for the decision to support a party and the party named. We show that most of those who display partisanship take cues from their parents, especially their mothers. Again, we begin with a Heckman Probit Selection model. Here too we examine the same related dependent variables: partisan support and preference.[5] This produces four models, one for each of the major parties in each country. And again, we employ the statistical program Stata/SE 8.2.

The results of our previous analyses guide this research. We include a variable that measures our primary theoretical concern: whether one or both of the parents names a party. We also provide a measure of partisan support in the respondent's region. Again, we expect the youngster's level of political interest to influence whether or not he or she names a party, but not the particular one chosen. We include several control variables: household income, union membership, socializing, voluntary and other social

---

[5] We continue to assume that children do not affect their parents; in the chapter's final models, we remove this assumption.

organizations, religious identification, gender, age (in order to observe whether exposure to a person's initial elections raises the probability of partisan support), and the year of the response (seeking to take account of the secular decline in partisanship). We add as well a measure of the national level of partisanship among young persons, in order to try to account for the possibility that general political issues influence a young person's partisan decision.

In order to account for the party named, we include measures of parental partisanship, the partisanship of the region, and the national level of preference for the party, as well as indicators of religion and social class (union membership in both countries and subjective social class identification in Britain), as well as perceptions of the economy, gender, age, and year. We expect the variables that assess parental partisanship and political interest to influence directly and strongly partisanship, no matter the impact of the more distant measures of social class and religion, and the importance of the young person's own level of political interest on the decision to support a political party. Our analysis links the young persons to the general process of bounded partisanship.

The results in Tables 5.4 and 5.5 display the determinants of partisanship in Germany, first for the SPD and then for CDU/CSU. Consider first partisan support. As we and many others have noted, a young person's level of political interest is a powerful predictor of the decision to support a party. So too is the presence in the young person's household of one or both parents who support a party. Partisan support in the region of residence also influences this response; other variables have minimal or no impact. Note especially that the national level of partisanship does not affect the dependent variable. We also find no evidence of a secular decline in partisan support during the years of the survey, after controlling for the other predictor variables. Partisan support responds to the youngster's interest in politics and to the cues offered by his or her parents; it does not derive from more distant structurally based changes in the party system. Parental party preferences display a very powerful impact on the child's selection of each of the major parties. Most of the other variables have little or no consistent impact on partisan choices. As expected, parental partisanship has a major impact on whether or not a child supports a party and the party chosen.

Applying postestimation techniques allows us to display in an intuitively meaningful manner the relative impact of parental partisanship on their children's partisan decisions. Here, we show two figures: one examines the determinants of partisan support and the second focuses on the choice of party. Both are products of the simulation of postestimation parameter estimates using the bootstrapping techniques that we describe in the Appendix.

Bounded Partisanship – Parents and Children

Table 5.4. *Choice of SPD among Young Persons in Germany – Heckman Probit Selection Model*

PANEL A: SPD CHOICE

	Coefficient	S.E.	Z-Score	Significance
Probit Model with Sample Selection				
Parents' party scale (left to right)	−0.26	.03	−9.9	.000
SPD choice in region, percentage	1.66	.69	2.4	.016
SPD choice in nation, percentage	4.39	1.72	2.6	.011
Union member	0.01	.08	0.1	.898
Volunteer activities	−0.01	.03	−0.2	.877
Social activities	0.00	.04	0.1	.919
No religious affiliation	0.01	.10	0.1	.929
Religious attendance	−0.07	.02	−3.0	.003
Household income	0.00	.00	−0.6	.571
Household income squared	0.00	.00	−0.5	.593
Worried about the economy	0.14	.04	3.3	.001
Age (16–29)	0.01	.01	2.2	.026
Female	0.31	.06	5.2	.000
Year	0.00	.01	−0.2	.816
Constant	1.68	15.38	0.1	.913

PANEL B: SUPPORT FOR ANY PARTY

	Coefficient	S.E.	Z-Score	Significance
Selection Model				
Parents support a party	0.17	.02	6.8	.000
Party support in region, percentage	−1.09	.51	−2.2	.032
Party support in nation, percentage	−0.16	.86	−0.2	.850
Political interest	0.32	.03	10.4	.000
Union member	0.05	.06	0.9	.396
Volunteer activities	−0.03	.03	−1.2	.244
Social activities	0.02	.03	0.9	.381
No religious affiliation	0.21	.09	2.4	.017
Household income	0.00	.00	−0.2	.843
Household income squared	0.00	.00	1.6	.120
Age (16–29)	0.00	.00	0.0	.984
Female	−0.22	.05	−4.6	.000
Year	0.00	.01	−0.4	.717
Constant	3.68	14.07	0.3	.793

$n = 3,114$
Log Likelihood = −2,706
Statistical probability of rho statistic =.000

Partisan Families

Table 5.5. *Choice of CDU/CSU among Young Persons in Germany – Heckman Probit Selection Model*

PANEL A: CDU/CSU CHOICE

	Coefficient	S.E.	Z-Score	Significance
Probit Model with Sample Selection				
Parents' party scale (left to right)	0.32	.04	8.4	.000
CDU/CSU choice in region, percentage	0.32	.83	0.4	.701
CDU/CSU choice in nation, percentage	3.24	1.52	2.1	.033
Union member	−0.19	.10	−1.9	.061
Volunteer activities	0.05	.04	1.5	.142
Social activities	−0.05	.04	−1.2	.250
Catholic	0.12	.07	1.9	.064
No religious affiliation	−0.28	.13	−2.2	.025
Religious attendance	0.03	.03	1.2	.231
Household income	0.00	.00	−0.7	.460
Household income squared	0.00	.00	0.2	.881
Worried about the economy	0.06	.05	1.2	.245
Age (16–29)	0.01	.01	1.2	.218
Female	−0.15	.09	−1.7	.083
Year	0.01	.01	1.0	.312
Constant	−18.39	17.36	−1.1	.290

PANEL B: SUPPORT FOR ANY PARTY

	Coefficient	S.E.	Z-Score	Significance
Selection Model				
Parents support a party	0.19	.03	7.0	.000
Party support in region, percentage	0.42	.55	0.8	.441
Party support in nation, percentage	−1.81	.89	−2.0	.042
Political interest	0.31	.03	9.8	.000
Union member	0.07	.07	1.1	.270
Volunteer activities	−0.01	.03	−0.5	.633
Social activities	0.02	.03	0.7	.465
No religious affiliation	0.21	.09	2.4	.017
Household income	0.00	.00	−0.5	.629
Household income squared	0.00	.00	1.9	.063
Age (16–29)	0.01	.00	1.6	.120
Female	−0.22	.05	−4.5	.000
Year	−0.01	.01	−1.5	.146
Constant	19.88	14.66	1.4	.175

$n = 3,008$
Log Likelihood $= -2,465$
Statistical probability of rho statistic $=.000$

Bounded Partisanship – Parents and Children

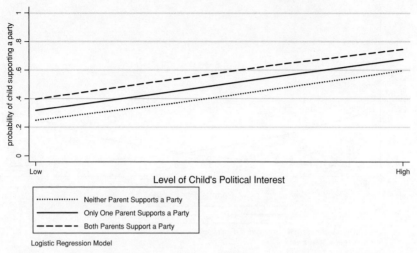

Figure 5.7. Postestimation Probabilities of Child's Partisan Support in Germany

In Figure 5.7, we illustrate our general argument by showing how variation in the parents' joint rate of supporting a party and the young person's level of political interest interact to affect the probability that he or she does so as well. The graph offers three trend lines: (1) neither parent supports a party; (2) one parent does; and (3) both mother and father pick a party. The x-axis includes the levels of the child's political interest, varying from none = 1 to high = 4. All other variables in the probit equation are held at their means. The predicted mean for the sample is 0.41, reminding us that on average young Germans do not select a party. The figure clearly displays the very powerful impact of the predictor variables on the child's decision to support a party. When he or she has no interest in politics and the parents do not name a party, the child is almost certain not to pick one. When he or she is very interested and both parents display partisanship, the predicted probability support is 0.8, twice the predicted mean. As the probit equation indicates, political interest is of greater importance than the parents' level of partisanship. In turn, as Tables 5.1 and 5.2 indicate, the young person's political interest responds to that of the parents, especially the mother. Of general importance, no matter the level of political interest, as one and then both the mother and father support a party, the rate of the child's support substantially increases.

Next, we apply the same technique to the choice of party. Here we emphasize one variable: the parents' partisan scale, which varies from −2, where both support the Social Democrats to +2, where both support the CDU/CSU, holding all the other variables at their means. Note

Partisan Families

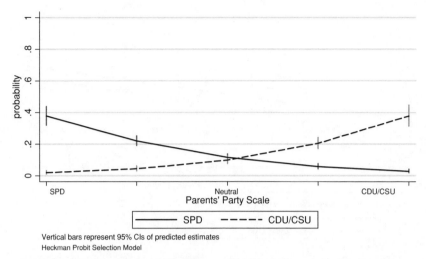

Figure 5.8. Postestimation Probabilities of Child's SPD and CDU/CSU Choice

that the two lines mirror each other: where both parents name the same major party, their offspring names that party at three times the predicted mean level (where the predicted mean is 0.10 for the SPD and 0.08 for the CDU/CSU) and almost never supports the other major party. It is important to underline that even in households in which both mother and father prefer the same major party, most young persons still are not expected to agree with them. No partisan choice remains the modal response for young Germans.

The British data reaffirm these patterns: the young person's level of political interest strongly affects partisan support, and parental partisanship powerfully influences whether or not the young person names a party and the party preferred. Table 5.6 presents the model for Labour, and Table 5.7 applies it to the Conservatives. The partisan distribution in the respondent's region influences both of the dependent variables, but the national level of partisanship does not. No matter the presence of better measures of social class (subjective social class, whether absent or present, as well as the particular class named, as well as union membership), parental partisanship is by far the most important determinant of party choice in Britain, as it is in Germany.

Again, we offer postestimation probabilities for the selection of a political party and for the party named. Here, the predicted probability of a child supporting a party, holding all predictors at their mean levels, is 0.50; British youngsters are as likely as not to pick a party, in an average year. The figure uses the same scales and dimensions as the analysis of the GSOEP data. Relative to the German data, the child's political interest

Bounded Partisanship – Parents and Children

Table 5.6. *Choice of Labour among Young Persons in Britain – Heckman Probit Selection Model*

PANEL A: LABOUR CHOICE

	Coefficient	S.E.	Z-Score	Significance
Probit Model with Sample Selection				
Parents' party scale (left to right)	−0.60	.05	−11.8	.000
Labour choice in region, percentage	2.47	.93	2.7	.008
Labour choice in nation, percentage	0.44	2.37	0.2	.852
Union member	0.19	.29	0.6	.529
Subjective social class	−0.15	.08	−1.9	.063
Church of England	−0.28	.20	−1.4	.155
No religious affiliation	0.21	.20	1.0	.312
Religious attendance	−0.07	.07	−1.1	.295
Household income	2.17	1.02	2.1	.034
Household income squared	−0.68	.44	−1.6	.121
Age (16–29)	0.00	.02	−0.2	.880
Year	0.01	.02	0.4	.692
Constant	−19.91	44.80	−0.4	.657

PANEL B: SUPPORT FOR ANY PARTY

	Coefficient	S.E.	Z-Score	Significance
Selection Model				
Parents support a party	0.16	.05	3.0	.002
Party support in region, percentage	−0.45	1.35	−0.3	.739
Party support in nation, percentage	−0.19	1.50	−0.1	.900
Political interest	0.64	.05	13.0	.000
Subjective social class dummy	0.16	.09	1.8	.077
Church of England	0.14	.11	1.3	.187
No religious affiliation	0.00	.12	0.0	.999
Religious attendance	0.01	.04	0.2	.851
Organizational memberships	−0.03	.04	−0.6	.536
Household income	−0.05	.41	−0.1	.904
Household income squared	0.04	.17	0.2	.832
Age (16–29)	0.03	.01	2.4	.018
Year	0.00	.03	0.2	.876
Constant	−10.35	53.43	−0.2	.846

$n = 1,217$
Log Likelihood = −918
Statistical probability of rho statistic = 0.77

has a greater impact on this choice. Here too the analysis displays the powerful joint influence of the young person's political interest and his or her parents' partisan decisions on these choices. Again, variation in partisanship of a child's mother and father affects the probability that he or she will name a party at each level of political interest.

Partisan Families

Table 5.7. *Choice of Conservatives among Young Persons in Britain –Heckman Probit Selection Model*

PANEL A: CONSERVATIVE CHOICE

	Coefficient	S.E.	Z-Score	Significance
Probit Model with Sample Selection				
Parents' party scale (left to right)	0.77	.07	10.3	.000
Conservative choice in region, percentage	−2.49	1.00	−2.5	.012
Conservative choice in nation, percentage	2.23	2.51	0.9	.376
Union member	0.21	.30	0.7	.488
Subjective social class	0.29	.08	3.5	.000
Church of England	0.58	.19	3.1	.002
No religious affiliation	−0.12	.22	−0.5	.590
Religious attendance	0.14	.07	1.9	.062
Household income	−1.72	1.12	−1.5	.123
Household income squared	0.59	.48	1.2	.224
Age (16–29)	0.03	.03	1.1	.255
Year	0.01	.02	0.4	.673
Constant	−20.84	47.56	−0.4	.661

PANEL B: SUPPORT FOR ANY PARTY

	Coefficient	S.E.	Z-Score	Significance
Selection Model				
Parents support a party	0.17	.05	3.3	.001
Party support in region, percentage	0.12	1.35	0.1	.929
Party support in nation, percentage	−0.25	1.49	−0.2	.866
Political interest	0.63	.05	12.5	.000
Subjective social class dummy	0.20	.09	2.3	.022
Organizational memberships	−0.02	.04	−0.5	.627
Church of England	0.13	.11	1.2	.225
No religious affiliation	0.00	.12	0.0	.976
Religious attendance	0.01	.04	0.4	.707
Household income	−0.04	.42	−0.1	.930
Household income squared	0.03	.17	0.2	.868
Age (16–29)	0.03	.01	2.3	.021
Year	0.00	.03	−0.1	.925
Constant	2.97	53.19	0.1	.955

$n = 1,217$
Log Likelihood = −873
Statistical probability of rho statistic = 0.52

Bootstrapping techniques enable us again to graph the impact of variation in parental support for a particular party on their child's partisan preference. Applying the results of Tables 5.6 and 5.7, Figures 5.8, 5.9 and 5.10 and shows the pronounced variation in the predicted probability of naming a party that follows variation in the partisan concurrence of

Bounded Partisanship – Parents and Children

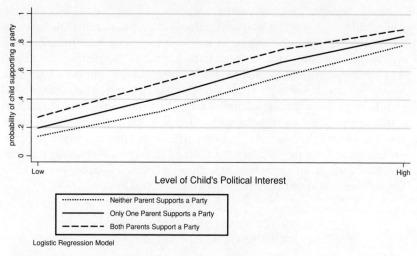

Figure 5.9. Postestimation Probabilities of Child's Partisan Support in Britain

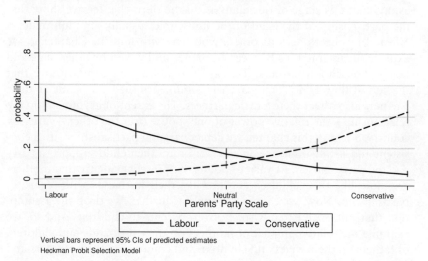

Figure 5.10. Postestimation Probabilities of Child's Labour and Conservative Choice

the parents. Where both support one of the major parties, the predicted probability of the young person also naming that party is 0.40 (where the mean predicted rate of naming Labour is 0.19 and the Tories is 0.06), and they are almost certain not to pick the other major party. In a political decade that saw major Labour gains and Tory losses, parental support further increases the probability of picking the victorious party as it also helps to insulate support for the losers from these general patterns.

Again, the two steps of partisanship are distinct but related choices. The child's own interest in politics and his or her parents' decision to name a party strongly influences the youngster's choice. Indeed, these two variables account for all the variation in the young people's decisions. Absent any political interest and lacking parents who support a party, the young person denies a partisanship. Having chosen a party, the second step follows from a more complex set of factors. Here, social class and religious variables, and the strength of the political parties in the region of residence affect the outcome, but they pale in comparison to the partisan preferences of the parents. Simply put and not at all surprising, parents' partisan choices strongly influence those of their children, even as the young persons do not mirror their parents' decisions.

RECIPROCAL PARTISANSHIP IN GERMAN AND BRITISH HOUSEHOLDS

Mothers, fathers, and their children compose the households that we examine at this stage of our analysis. In the third chapter, we show that the partisanship of the head of the household, usually an adult male, is affected by the choices of others who live with him. In Chapter 4, we explore the interactions between husbands and wives, or other male and female household partners. This analysis of couples demonstrates that the causal flow is not one way; wives influence their husbands as male partners also affect their female partners. The extent of asymmetry in the relationship – the relative power of one spouse over the other – varies by country. So far in this chapter, we depict parental partisanship with a consistent and strong influence on whether or not their child supports a party and the party chosen. At various points, we present evidence that mothers are more like their children and less like their husbands with regard to partisanship. Now, we extend the analysis further. We drop the assumption that influence extends solely from parents to children. And so, we examine the relative impact of mothers on their husbands and children, of fathers on their wives and children, and children on their mothers and fathers. The results consistently point to the centrality of wives/mothers in the flow of partisan influence within German and British households.[6]

In principle, we expect wives/mothers, husbands/fathers, and children to influence each other regarding the dimensions of partisanship and

---

6 And see Acock and Bengtson (1978) for a classic analysis that emphasizes the relative importance of the mother in the transmission of political values. Smetana (2005) introduces papers that review recent research in adolescent psychology, addressing reciprocal relations within families, and see Steinberg and Silk (2002). Oygard et al. (1995) compares in effects of parents and peers on smoking.

Bounded Partisanship – Parents and Children

almost everything else. Living together and trusting each other induces sending and receiving partisan cues. There is, however, no reason to expect that these relationships are always present and are always reciprocal. In the previous chapter, we showed that husbands and wives always influence each other, but their relative influence varies across the two countries. Here, we present evidence that mothers always influence their children and their husbands; children usually influence their mothers, but not their fathers, and husbands/fathers influence their wives but usually not their children. Also, where two members of the household share the same partisan preference, they have a very strong influence on the third, and this applies to each of the three relationships: mothers and fathers affecting their child; wives/mothers and child influencing the husbands/fathers, and parent and child affecting the other parent. Although husbands/fathers are generally the most interested in politics, they do not dominate the picture of households that we draw with the data from GSOEP and BHPS. In the next chapter, we show that wives/mothers are also central to electoral decisions in Britain (GSOEP lacks the relevant data). In the concluding chapter, we account for the centrality of mothers/wives in the flow of political influence within households, and we suggest how their pivotal role might serve as a fulcrum of electoral change.

In order to sketch these results, we use three-stage least squares estimation for systems of simultaneous equations, in the form of a linear probability model. This allows us to estimate statistically the reciprocal effect of each member on the others.[7] From these results, we derive the predicted probabilities, which enable us to depict the flow of influence within households regarding partisan preferences.

We restrict our analysis to three categories of persons: (a) children aged 16–29 of mothers and fathers in the sample, (b) wives/mothers who live with their male partner and have at least one child aged 16–29 in the sample, *and* (c) husbands/fathers who live with their female partner and have at least one child aged 16–29 in the sample. The respondents are members of the same families, forming mother–father–child triads. Because

---

7 Heckman and Macurdy (1985) and Heckman and Snyder (1997) justify the application of least squares regression analysis to a bifurcated dependent variable. Instrumental variable probit analysis does not permit the analysis of three-way reciprocal effects. Upon examination, the outcomes of a series of two-stage instrumental probit analyses substantially mimic those obtained from the three-stage OLS regression model. We ran a series of bootstrap simulations, using the parameter estimates from these models to determine the predicted probabilities of each dependent variable. These use the complete range of values on the predictor variables. The 95 percent confidence intervals of the point estimates of the predicted probabilities always fall between zero and one. Again, we use Stata/SE 8.2 to predict linear estimates. The results are available from the authors.

Table 5.8. *Agreement on SPD Choice in Germany among Mothers, Fathers, and Children – Three-Stage Regression Model*

	Dependent Variable											
	Mother Chooses SPD				Father Chooses SPD				Child Chooses SPD			
	Coefficient	S.E.	Z-score	Significance	Coefficient	S.E.	Z-score	Significance	Coefficient	S.E.	Z-score	Significance
Mother chooses SPD					0.17	.05	3.4	.001	0.23	.04	5.8	.000
Father chooses SPD	0.52	.03	15.6	.000					0.24	.04	6.1	.000
Child chooses SPD	0.23	.05	4.5	.000	0.15	.06	2.4	.015				
Mother's political interest	0.09	.00	18.2	.000								
Father's political interest					0.06	.01	10.8	.000				
Child's political interest									0.06	.00	14.5	.000
Mother union membership	0.08	.01	6.6	.000								
Father union membership					0.14	.01	14.9	.000				
Child union membership									0.04	.01	4.6	.000
Occupation:												
Mother self-employed w/co-workers	0.03	.01	3.6	.000								
Father self-employed w/co-workers					0.02	.01	1.1	.254				

	β	SE	z	p	β	SE	z	p	β	SE	z	p
Child self-employed w/co-workers	−0.01	.02	−0.4	.658					0.01	.01	1.2	.222
Mother self-employed w/no co-workers												
Father self-employed w/no co-workers					−0.11	.01	−8.0	.000	−0.03	.03	−0.9	.345
Child self-employed w/no co-workers	0.07	.02	4.3	.000								
Mother skilled manual												
Father skilled manual					0.08	.01	8.5	.000	−0.01	.01	−1.5	.144
Child skilled manual	0.01	.01	1.5	.138								
Mother unskilled manual												
Father unskilled manual					0.08	.01	7.4	.000				
Child unskilled manual									0.01	.01	1.5	.129
Age (16–29)					0.02	.01	1.5	.147	0.01	.00	6.0	.000
Constant	−0.13	.01	−11.1	.000					−0.23	.02	−10.2	.000
	$n = 11{,}499$				$n = 11{,}499$				$n = 11{,}499$			
	$R^2 = .44$				$R^2 = .28$				$R^2 = .18$			
	$\text{Chi}^2 = 1{,}545, p = .000$				$\text{Chi}^2 = 1{,}194, p = .000$				$\text{Chi}^2 = 810, p = .000$			

Endogenous variables: mother's SPD choice, father's SPD choice, child's SPD choice
Instrumental variables: occupation measures, union membership, political interest (for mothers, fathers, and children)

Table 5.9. Agreement on CDU/CSU Choice in Germany among Mothers, Fathers, and Children – Three-Stage Regression Model

	Dependent Variable											
	Mother Chooses CDU/CSU				Father Chooses CDU/CSU				Child Chooses CDU/CSU			
	Coefficient	S.E.	Z-score	Significance	Coefficient	S.E.	Z-score	Significance	Coefficient	S.E.	Z-score	Significance
Mother chooses CDU/CSU					0.38	.08	4.5	.000	0.39	.08	5.1	.000
Father chooses CDU/CSU	0.61	.04	15.9	.000					0.07	.06	1.1	.259
Child chooses CDU/CSU	0.21	.06	3.9	.000	0.02	.07	0.3	.737				
Mother's political interest	0.03	.00	7.0	.000								
Father's political interest					0.04	.01	7.8	.000				
Child's political interest									0.06	.00	15.5	.000
Mother union membership	−0.06	.01	−4.9	.000								
Father union membership					−0.10	.01	−9.3	.000				
Child union membership									−0.03	.01	−3.0	.003
Occupation: Mother self-employed w/co-workers	−0.01	.01	−1.6	.107								

	Mother's CDU/CSU choice				Father's CDU/CSU choice				Child's CDU/CSU choice			
Father self-employed w/co-workers	0.04	.02	2.2	.028	−0.02	.01	−1.3	.188	0.00	.01	0.0	.985
Child self-employed w/co-workers												
Mother self-employed w/no co-workers												
Father self-employed w/no co-workers	−0.04	.02	−2.5	.014	0.09	.02	6.1	.000	0.03	.03	1.1	.284
Child self-employed w/no co-workers												
Mother skilled manual	0.01	.01	0.6	.558	−0.06	.01	−5.7	.000	0.02	.01	2.6	.010
Father skilled manual												
Child skilled manual												
Mother unskilled manual	−0.01	.01	−0.8	.410	−0.08	.01	−7.5	.000	0.01	.01	1.4	.164
Father unskilled manual												
Child unskilled manual												
Age (16–29)									0.00	.00	5.1	.000
Constant					0.15	.02	7.2	.000	−0.23	.02	−10.6	.000
	$n = 11{,}499$				$n = 11{,}499$				$n = 11{,}499$			
	$R^2 = .41$				$R^2 = .35$				$R^2 = .18$			
	$\text{Chi}^2 = 966, p = .000$				$\text{Chi}^2 = 1{,}084, p = .000$				$\text{Chi}^2 = 759, p = .000$			

Endogenous variables: mother's CDU/CSU choice, father's CDU/CSU choice, child's CDU/CSU choice
Instrumental variables: occupation measures, union membership, political interest (for mothers, fathers, and children)

they almost always live together during the years in which they respond to the survey's questions, we use both family and household to describe the social relationship.

We model each person's party choice as the result of two sets of variables: (a) his or her own personal characteristics, namely measures of social class and political interest (because of its indirect impact via partisan support) and (b) instrumental variables for the others in the household – measures that correlate with each person's partisan preference but not with anyone else's choice or any of the predictors of that decision. The instrument is defined by political interest and social class (union membership in both Germany and Britain and various occupations in Germany and subjective social class in Britain), but not religious identification, because of its relatively high polychoric correlation across the members of the family (see Tables 5.1–5.2). Like our previous models (indeed all models), this analysis has obvious strengths – a direct analysis of reciprocity – but it also has limitations – we can only address one dependent variable at a time; we cannot examine partisan constancy, and we cannot control for clustering or autocorrelation. At this stage of our analysis, we focus all our attention on the effect of the partisan preferences of others in the household on the choice of party.

Table 5.8 applies the model to the choice of the Social Democrats. No matter the importance of a person's level of political interest and union membership (and they are always substantial), each member of the household influences every other member. Table 5.9 presents the results for the German Christian Democratic Party/Christian Social Union. In Germany, nine of the twelve strands that link members of the household show statistically significant effects (the exception is children and their fathers regarding the CDU/CSU preference). Again, political interest and union membership influence partisan preference as well. The Z-scores indicate that the husband's impact on his wife emerges as the strongest of the dyadic relationships. These simple models offer complex and affirming results.

Next, we model the reciprocal nature of partisan choices in British households. We use the same instrumental variables: political interest and social class (union and subjective class identification, but not religion), and again we control for each respondent's level of political interest and social class characteristics. Some strands display no effects of any kind: no exchange occurs between fathers and their children with regard to Labour support, after controlling for the other variables in the equation. We also see a negative relationship between children and their fathers with regard to the Tories, and here too fathers do not influence their children. Only the wives/mothers consistently influence their husbands and children. Table 5.10 models Labour preference, and Table 5.11 displays the relationship with regard to Tory support.

Table 5.10. Agreement on Labour Choice in Britain among Mothers, Fathers and Children – Three-Stage Regression Model

	Dependent Variable											
	Mother Chooses Labour				Father Chooses Labour				Child Chooses Labour			
	Coefficient	S.E.	Z-score	Significance	Coefficient	S.E.	Z-score	Significance	Coefficient	S.E.	Z-score	Significance
Mother chooses Labour					0.68	.10	6.5	.000	0.60	.12	4.9	.000
Father chooses Labour	0.48	.05	8.9	.000					0.04	.09	0.4	.681
Child chooses Labour	0.20	.06	3.4	.001	0.02	.07	0.2	.827				
Mother's political interest	0.01	.01	2.1	.033								
Father's political interest					0.07	.01	8.4	.000				
Child's political interest	0.05	.01	4.1	.000					0.09	.01	12.6	.000
Mother union membership												
Father union membership					0.07	.01	5.4	.000				
Child union membership									0.08	.02	4.1	.000
Mother's subjective social class	−0.04	.01	−5.6	.000								
Father's subjective social class					−0.05	.01	−7.4	.000				
Child's subjective social class									−0.03	.01	−4.0	.000
Age (16–29)									0.01	.00	2.9	.004
Constant	0.06	.02	2.8	.005	−0.05	.02	−2.0	.044	−0.25	.04	−6.3	.000
	$n = 4{,}352$				$n = 4{,}352$				$n = 4{,}352$			
	$R^2 = .37$				$R^2 = .34$				$R^2 = .16$			
	$\text{Chi}^2 = 477, p = .000$				$\text{Chi}^2 = 690, p = .000$				$\text{Chi}^2 = 565, p = .000$			

Endogenous variables: mother's Labour choice, father's Labour choice, child's Labour choice
Instrumental variables: union membership, subjective social class (for mothers, fathers and children)

Table 5.11. *Agreement on Conservative Choice in Britain among Mothers, Fathers and Children – Three-Stage Regression Model*

	Dependent Variable											
	Mother Chooses Conservatives				Father Chooses Conservatives				Child Chooses Conservatives			
	Coefficient	S.E.	Z-score	Significance	Coefficient	S.E.	Z-score	Significance	Coefficient	S.E.	Z-score	Significance
Mother chooses Conservatives					0.77	.08	9.0	.000	0.58	.09	6.6	.000
Father chooses Conservatives	0.45	.06	6.9	.000					0.07	.08	0.9	.384
Child chooses Conservatives	0.11	.08	1.4	.175	−0.29	.09	−3.1	.002				
Mother's political interest	0.04	.01	5.6	.000								
Father's political interest					0.02	.01	2.4	.015				
Child's political interest									0.07	.01	11.5	.000
Mother union membership	−0.08	.01	−6.0	.000								
Father union membership					−0.05	.01	−3.8	.000				
Child union membership									−0.03	.02	−1.8	.076
Mother's subjective social class	0.04	.01	5.8	.000								
Father's subjective social class					0.06	.01	8.4	.000				
Child's subjective social class									0.02	.01	3.1	.002
Age (16–29)									0.00	.00	2.7	.007
Constant	0.03	.02	1.8	.068	0.10	.03	3.8	.000	−0.24	.04	−6.2	.000
	$n = 4,355$				$n = 4,355$				$n = 4,355$			
	$R^2 = .34$				$R^2 = .19$				$R^2 = .09$			
	$\text{Chi}^2 = 434, p = .000$				$\text{Chi}^2 = 389, p = .000$				$\text{Chi}^2 = 440, p = .000$			

Endogenous variables: mother's Conservative choice, father's Conservative choice, child's Conservative choice
Instrumental variables: union membership, subjective social class (for mothers, fathers, and children)

Bounded Partisanship – Parents and Children

Table 5.12. *Predicted Probabilities of Partisan Choice*

	Germany		Britain	
Household Support	SPD	CDU/CSU	Labour	Conservatives
Wife/Mother				
Mean	0.31	0.25	0.32	0.30
Husband and child	0.83	0.88	0.78	0.68
Husband not child	0.60	0.66	0.58	0.57
Child not husband	0.31	0.26	0.30	0.23
Neither	0.08	0.05	0.10	0.12
Husband/Father				
Mean	0.30	0.32	0.34	0.32
Wife and child	0.54	0.62	0.86	0.60
Wife not child	0.39	0.60	0.84	0.90
Child not wife	0.37	0.24	0.18	<.01
Neither	0.22	0.21	0.16	0.13
Child				
Mean	0.19	0.14	0.22	0.21
Mother and father	0.49	0.48	0.69	0.63
Father not mother	0.26	0.08	0.09	0.06
Mother not father	0.26	0.41	0.65	0.56
Neither	0.02	0.02	0.05	<.01

Source: Results of Tables 5.10 and 5.11

Simulating postestimation probabilities adds nuance and deepens our analysis. We present the results for all the flows of influence for the choice of each of the parties in Table 5.12. In all cases, where neither the mother nor the father selects Party A/B, the child is certain not to name the party, with all the other predictor variables at the means. Absent the joint influence of both parents, partisan choice does not occur. Where both mother and father name the party, the predicted probability that the child will do so as well is 2.5 to 3 times the predicted mean. Where only one parent picks the party, the dominance of the mother's influence emerges, standing as much as ten times that of the father (for the Conservative Party). Children's partisan choice responds to their mother's influence and to that of their parents combined, and least to their father's preferences.

Note too the extent to which the responses of the wife/mother parallel that of their children. Absent husband and child who select Party A/B, she is just about certain not to name the party (with a predicted probability of selection around 0.10, holding all the other predictors at the means). Where both pick the party, the probability of the mother doing so too doubles and triples the predicted means. Here, however, the husband's effect on the wife far exceeds that of the child, but it is important to notice that the child has an additional positive impact on the mother's partisan choice, where the husband/father supports that party.

Husbands/fathers stand somewhat apart from these patterns. First, the influence and reception of the others' preferences is different in the two countries. In Germany, though strong, household effects are weaker for the choice of the SPD than the CDU/CSU. Note additionally that the child's impact on the father's predicted probability of picking the Social Democrats is not much different from that of the mother. In Britain, fathers and children have no positive impact on each other at all. Indeed, the child's choice of a party dampens the predicted probability that the father does so at a level below the predicted mean. Still here too the combined effect of the other family members always doubles the predicted probability of partisan choice, and again the wife's selection of Party A/B doubles the predicted probability that her husband does so as well.

Consider now how household members in combination affect each other. Again wives/mothers are central. Their husbands and children acting in concert have a greater impact than each taken alone. When both name Party A, the wife's/mother's predicted probability of support varies from two to four times the mean. Husbands/fathers are different; the effect of children does not usually add to the wives'/mothers' impact.[8] In three cases, mothers are so important to the children's choice that adding the father's choice affects concurrence by less than .10. Note also that where two household members do not support the same party, the probability of partisan choice for the other drops well below the mean (ranging from less than .05 of the mean for children and the Conservatives to .67 for husbands/fathers and the two German parties). For mothers/wives and children, the absence of concurring support from the others in the household makes it virtually certain that they will not name Party A/B. Households have cumulative effects, even as individual members influence each other's partisan preference.

It follows from this analysis that reciprocal effects within households appear as well with regard to the *absence* of partisan support. We repeat the same linear probability model, with no party choice as the dependent variable. Here too each member influences the others, no matter the impact of the control variables (age and political interest for the children and political interest for the parents). Rather than present another set of tables, we summarize the results. In Germany, wives/mothers have the highest level of impact, and both their husbands and children strongly influence their own decisions not to support a party. Again in Britain,

---

8 Indeed, in the case of the Conservatives, it seems to weaken it, an outcome that we expect derives from the small numbers who make this choice and the instability in the data.

Bounded Partisanship – Parents and Children

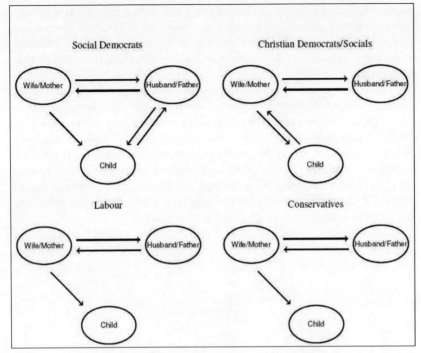

Figure 5.11. Political Influence in Households

wives/mothers influence and are influenced by everyone else in the family. Here, they have slightly less impact on their child's lack of partisan support than do the husbands/fathers, but here too the children do not influence their fathers at all.

The four panels of Figure 5.11 provide an easily recognized picture of the flow of partisan influence within German and British households. In both of these established democracies, the wife/mother stands at the center of family politics. She always influences the husband's and children's partisan choices. In turn, the husband always affects her, and in conjunction with the husband/father her offspring do so in every case, including naming no party. The husband/father has a more complex position in the household. In Germany, he influences the wife (indeed more than she affects him), but he has less impact on the children's partisan choice or their decision to name no party. In Britain, he has a negative impact on the predicted probability that the child supports the Tories, and the offspring do not influence his decision not to support a party. Children, we see, are always affected by their mothers and sometimes by their fathers, and they sometimes influence one or both of their parents, especially when they are joined by the other parent. This finding reaffirms the decision to model the partisan choice of children as a result of a reciprocal relationship with their parents, and not as a decision that only depends on what their parents do. Wives/mothers stand at the center of reciprocal partisan households. In the conclusion of the volume, we explore the bases of this relationship.

In the next chapter, we extend the analysis to voting, a small but important step beyond partisan preference. We relate electoral behavior – the defining act of democratic citizenship – to a person's past partisan choices and to the voting decisions of members of the household. Turnout and vote choice respond directly to variations in party support in the years prior to an election and to the voting choices of wives/mothers, husbands/fathers, and children. In the concluding chapter, we discuss the theoretical implications of our analysis and examine the relationship between bounded partisanship and party politics at the national level.

6

Partisan Constancy and Partisan Families: Turnout and Vote Choice in Recent British Elections

> Man may not be a political animal, but he is certainly a social animal. Voters do respond to the cues of commentators and campaigners, but only when they can match those cues up with the buzz of their own social group. Individual voters are not rational calculators of self-interest (nobody truly is), and may not be very consistent users of heuristic shortcuts, either. But they are not just random particles bouncing off the walls of the voting booths. Voters go into the booth carrying the imprint of the hopes and fears, the prejudices and assumptions of their family, their friends, and their neighbors. For most people, voting may be more meaningful and more understandable as a social act than as a political act.
> Louis Menand, *The New Yorker*, August 30, 2004, 96

We have developed a social logic of partisanship: the partisan choices of heads of households, at one point in time and over time, are affected by the partisan preferences of others in the family, as well as other more distant and abstract social locations; the level of partisan similarity between husbands and wives and parents and children reflects their shared social locations, and the bounded nature of partisanship; and the length of marriage influences partisan concurrence in couples. Spouses influence each other. There are as well reciprocal effects among the partisan choices of household partners and parents and children. Mothers/wives influence their husbands and children; the partisan preferences of children sometimes affect mothers and fathers; husbands/fathers influence their wives, but usually not their children. We have explored three elements of the concept partisanship: partisan support (whether or not a person names a party), partisan preference (the party chosen), and partisan constancy (the rate by which a person picks a party). Our analysis finds that most Germans and Britons are bounded partisans: at a single point in time, most pick a party; over time, hardly any ever pick both of the dominant parties; instead they vary their choices between one of the major parties

Partisan Families

and no party. We find that living together, bounded partisans sustain each other.

Now, we draw these themes together into an analysis of voting behavior, casting a ballot for the candidate, candidates, or list that represents a political party. We add a new dimension of partisanship, support constancy (the rate by which a person names any party over time), and so we distinguish here between this variable and preference or choice constancy (the rate of picking the same party). The analysis shows that electoral behavior responds to both elements of partisan constancy and the voting decisions of others in the household. Support constancy affects turnout, and choice constancy influences the decision at the ballot box, net of other predictors. The relationship between partisanship and vote choice is close, but not tautological.[1] Furthermore, we show that living with others influences turnout, no matter a person's level of political interest, age, health, and host of other variables. In addition, electoral decisions of wives/mothers, husbands/fathers, and children combine with partisan constancy to explain vote choice. Recall too that partisanship – support, preference, and constancy – respond to the influence of intimate and more distant and abstract social ties. As the social observer Louis Menand understands, electoral decisions reflect a social logic.

Of the two surveys that provide most of our data, only BHPS includes questions on electoral decisions, and so this chapter examines behavior only in British elections. The questions that tap electoral choice are straightforward. BHPS asks the respondents whether or not they voted in the election of that year, and, if so, for which party. In this analysis too, we focus on Labour and the Conservatives.

We apply the same models that we use in the previous chapters. We analyze turnout and vote choice with a Heckman Probit Selection model. Here, we include the constancy of partisan support in the years prior to the election as well as a host of factors that characterize people's social and personal lives, but not the effects of others in the household on their voting decisions. We then apply a three-stage least squares linear probability model using instrumental variables to explore the reciprocal effects on vote choice in British households. This examines voting decisions, net of the effects of partisan constancy and measures of social class and religion. Once again, the analyses use Stata SE/8.2.

---

[1] Beginning with Butler and Stokes's analysis of British politics (1969; 1974), the relevant literature recognizes the close ties between partisanship and voting; indeed, some scholars insist that the relationship is a tautology. For a recent example using BHPS, see Brynin and Sanders (1997). In the Preface, we explain why we do not accept this claim.

Turnout and Vote Choice in Recent British Elections

Though we emphasize the importance of intimate social ties on political behavior, we do not ignore the political world outside the household. During the years of our analysis, British politics is transformed. In 1997, Labour replaces the long governing Conservatives, as Tony Blair and his party remove John Major from the prime minister's office and turn the Conservative Party into an ineffective opposition. The following election in 2001 deepens Labour's control over the electorate and the government. Our analysis includes, therefore, variables that measure each party's strength at the regional and national levels in a given year. These serve as additional controls for our claim that electoral choice responds to intimate social ties.

We model turnout and vote choice with a limited set of variables. Focusing primarily on data from BHPS, our analysis can pay no attention to the various attitudes that accompany many, indeed, most studies of electoral behavior: views of the candidates, general political attitudes, questions about views on the economy (which we include only in the base-line models of Chapter 2). We stay with the social context of partisanship and voting decisions, avoiding the difficulty of separating these attitudes from our dependent variables (Anderson, Mendes, and Tverdova 2004; Erikson 2004; Johnston et al. 2005). We are more concerned by our inability to include measures of the political parties as agents that bring people to the polls and that influence vote choice. Party strength in the region and nation are at best distant indicators of these critical phenomena. We would prefer to be able to include information that taps various elements of election campaigns – television broadcasts, print advertisements, door-to-door efforts, telephone campaigns, and the like. Absent these measures, we can focus only on the proximate causes of electoral decisions, as they play out within households. In the next and concluding chapter, we return to this theme, by adding the role of political parties to a discursive, but not statistical, effort to model the interaction between politics in households and national politics.

THE SOCIAL LOGIC OF ELECTORAL CHOICE IN THE BRITISH GENERAL ELECTIONS OF 1997 AND 2001

Voting entails two related choices. At analytically distinct moments, people decide whether or not to cast a ballot and for which party to vote. Each of these decisions may be conceived as a binary dependent variable. Again, we apply the Heckman Probit Selection model to address this two-stage process. Recall that this model has several strengths: it answers these questions sequentially in the same model, and it offers a summary statistic that describes the strength of association between the answer to the first and second questions. The analysis uses the cross-sectional responses to

Partisan Families

draw from the respondents' answers, presenting the results for an "average election."[2] There are 10,743 observations; these combine persons and their responses in the election year and the two preceding years (0.65 of the observations apply to persons who live with others, and 0.35, to those who live alone). We construct different models for each party. All our analyses use the statistical program Stata/SE 8.2.

Each portion of the table adds to the analysis. In the bottom half, we model turnout as a function of a set of predictor variables. Here, we are especially concerned to show that living alone or not influences this decision, controlling for the effects of the other variables.[3] The top half provides the base-line variables for the analysis of vote choice that we include in our more complex analysis of reciprocal effects.

We model turnout as the function of two sets of variables. One cluster includes variables that are well known to influence the decision to vote or not: age cohorts (contrasting those between the ages of 30 and 50 and persons older than 50 with the youngsters); political interest; education; the frequency by which a person supports a political party in the two years preceding the election;[4] dummy variables for social class identification, religious identification, and organizational membership; and a measure of partisan support in the respondent's region of residence and one for the nation as a whole. The second cluster adds variables that may simultaneously influence the probability of living alone and turnout; a composite measure of subjective well-being; a subjective assessment of

---

[2] Because political parties and electorates change (as they do during the years we study Britain), we also performed separate analyses for each of the three election years covered. Our results demonstrate no significant change in the relative importance of predictor variables in determining electoral vote from 1992 to 2001.

[3] Using data drawn from surveys in southern California from the 2000 national elections, Gray (2003) also finds that living with others affects turnout in the primaries preceding the election and in the presidential election itself. The analysis shows powerful effects of spouse's turnout on their mate's decision to vote, net of a host of predictor variables. "The results of Table 4 show that where more of the registered housemates voted and or when the respondent is married and a spouse cast a ballot it is much more likely that they would have voted as well. Controlling for all other factors in the model, a respondent living in a home where all other registered housemates vote is about four times more likely than a respondent living in a home where no other registered members vote." Gray finds the same effect for spouses on the respondents and notes that this impact is independent of overt "mobilization by other housemates" (2003, 40). Petersson, Westholm, and Blomberg (1989) also locate lower rates of political participation among persons who live alone in Sweden.

[4] Looking at the two years prior to an election allows us to maintain a large sample, while detailing variation. Because the surveys begin in the year before the 1992 general elections, we exclude those responses from the analysis, and so we focus on turnout and vote choice in 1997 and 2001.

the respondent's health during the past year; a subjective assessment of the ability to do one's daily activities; and the frequency that the respondent meets persons outside the household each week. These along with the measures of age allow us to control for the possibility that questions of health, the ability to get around, and social interactions influence both living alone and turnout. This provides a demanding test for our hypothesis that simply living with others or alone influences the probability of casting a ballot.

The top portion of the model introduces the variables that affect vote choice: partisan constancy for the party; subjective social class (on a scale that varies from working class to middle class and associates the former with Labour votes and the latter with the Tories); identification with the Church of England (with the expectation of a negative impact on choosing Labour and positive for the Conservatives); union membership (positive for Labour and negative for the Conservatives); and partisan strength in the region of residence and the nation. Because living alone or with others is not expected to and does not affect vote choice, we omit it from this portion of the analysis. This model serves as a base-line, against which we measure the impact of how others in the household vote, in the next section. Table 6.1 presents the determinants of turnout and vote choice with regard to Labour, combining responses for the 1997 and 2001 general elections. Table 6.2 repeats the analysis for the Conservatives.

These tables demonstrate that the probability of turnout is directly affected by the intimate social context of a person's life: living alone reduces the level, even after controlling for all the other predictor variables. It reaffirms as well the importance of partisan constancy, as well as the level of partisanship in the region and nation, and it sustains the well-known importance of more general social identifications, memberships, and interactions. Also, the rho statistic shows that the two choices – whether or not to vote and for which party – are related.

These results support Fowler's (2005) effort to model turnout as related to the behavior of members of a person's social network (and see Beck et al. 2002; Knack 1994; Mutz and Mondak 1997; and Straits 1990, as well as the work that draws directly on the Columbia School of electoral analysis, which we describe in Chapter 1). Because three fourths of the Britons report that they vote, living with others and meeting with them in various social locations necessarily includes interactions and exchanging cues with persons who are likely to vote. As the first step in our analysis, we note that the evidence taken from BHPS supports the claim that the decision on turnout derives from a social process.

What explains vote choice? Drawing on these tables, we note that partisan constancy is a powerful predictor of vote choice, offering by far the highest Z-scores. As has long been known as well, subjective identification with the middle or working class, identification with the Church of

Partisan Families

Table 6.1. *Labour Vote – Heckman Probit Selection Model*

PANEL A: VOTES LABOUR

	Coefficient	S.E.	Z-Score	Significance
Probit Model with Sample Selection				
Chooses Labour past two years (0.2)	1.11	.02	52.4	.000
Labour preference in region, percentage	1.31	.17	7.6	.000
Labour preference in nation, percentage	3.93	.92	4.3	.000
Union membership	0.04	.04	1.1	.292
Subjective social class	−0.12	.02	−7.4	.000
Church of England member	−0.08	.03	−2.6	.010
Constant	−2.83	.32	−9.0	.000

PANEL B: VOTES

	Coefficient	S.E.	Z-Score	Significance
Selection Model				
Party support last two years (0.2)	0.48	.02	25.7	.000
Party support in region, percentage	0.86	.43	2.0	.044
Party support in nation, percentage	2.95	.57	5.2	.000
Political interest	0.26	.02	14.1	.000
Has a subjective social class	0.16	.04	3.9	.000
Organizational memberships	0.08	.03	2.7	.007
How often meets people	0.00	.02	0.2	.868
Lives alone	−0.23	.04	−6.1	.000
Has a religion	0.11	.03	3.7	.000
Education (ref: no degree)				
University degree	0.24	.06	3.8	.000
Upper school qualification	0.13	.04	3.2	.001
Lower school qualification	0.06	.04	1.6	.114
Subjective well-being	0.00	.01	1.0	.338
Health status	−0.03	.02	−1.5	.134
Health limitations	0.04	.05	0.9	.353
Middle aged (31–50)	0.33	.03	9.7	.000
Older than 50	0.57	.04	12.6	.000
Constant	1.27	.20	6.4	.000

$n = 10{,}743$
Log likelihood $= -7{,}692$
Statistical probability of rho statistic $= 0.00$

England, and union membership affect electoral behavior, as does the party's success in the nation and in the respondent's region in that year.[5]

[5] It is worth highlighting that social contexts predict vote choice very well, see Goldthorpe (1999a) and Johnston and Pattie (2005). Similarly, Johnston (1999)

Turnout and Vote Choice in Recent British Elections

Table 6.2. *Conservative Vote – Heckman Probit Selection Model*

PANEL A: VOTES CONSERVATIVE

	Coefficient	S.E.	Z-Score	Significance
Probit Model with Sample Selection				
Chooses Conservatives past two years (0.2)	1.33	.02	54.1	.000
Conservative preference in region	2.13	.35	6.1	.000
Conservative preference in nation	−2.69	1.31	−2.1	.039
Union membership	−0.17	.05	−3.4	.001
Subjective social class	0.09	.02	4.4	.000
Church of England member	0.12	.04	3.3	.001
Constant	−1.51	.23	−6.5	.000

PANEL B: VOTES

	Coefficient	S.E.	Z-Score	Significance
Selection Model				
Party supports last two years (0.2)	0.44	.02	23.2	.000
Party support in region, percentage	0.33	.43	0.8	.447
Party support in nation, percentage	3.49	.58	6.0	.000
Political interest	0.25	.02	12.8	.000
Has a subjective social class	0.19	.04	4.4	.000
Organizational memberships	0.09	.03	2.9	.004
How often meets people	0.03	.02	1.7	.084
Lives alone	−0.26	.04	−6.4	.000
Has a religion	0.15	.03	4.9	.000
Education (ref: no degree)				
University degree	0.15	.07	2.2	.028
Upper school degree	0.05	.04	1.1	.276
Lower school degree	0.01	.04	0.1	.895
Subjective well-being	0.00	.01	0.9	.358
Health status	−0.01	.02	−0.7	.498
Health limitations	−0.01	.05	−0.2	.864
Middle aged (31–50)	0.32	.04	9.0	.000
Older than 50	0.56	.05	12.0	.000
Constant	1.25	.21	6.1	.000

$n = 10.743$
Log likelihood $= -6,932$
Statistical probability of rho statistic $= 0.00$

The successes of these other variables also underline the analytical difference between partisan support and voting; one does not define the other. The variables that describe social characteristics and the distribution of

and Beck et al. (2002) note that personal discussion networks play an important role in explaining electoral choices that differ from partisanship.

political preferences in the nation and region serve as controls and assist our ability to create instrumental variables for the forthcoming analysis, which details variations in the reciprocal nature of political influence within households.

ASSESSING THE RECIPROCAL IMPACT OF PARENTS AND CHILDREN ON EACH OTHER'S VOTING DECISIONS

Electoral choice for any one person, we hypothesize, is a joint function of a person's partisan constancy and the electoral decisions of other members of their household. As in the previous chapter, our analysis draws particular attention to the importance of wives/mothers in this process. Mothers wives always affect their husbands and children; husbands fathers always affect their wives, but do not always influence their children; and children sometimes influence their parents.

As we discuss in Chapter 5, modeling this social relationship encounters an obvious problem: it is by definition an endogenous relationship. We expect A, B, and C, who live together to influence each other, with regard to electoral choice and almost everything else. The strength of any one strand in a dyad, indeed whether or not it is present at a particular moment in a family or set of families, is an empirical question, which can only be answered by parsing the relative effects of A, B, and C on each other. Here too our solution applies the logic of the linear probability model (instrumental variables in a three-stage least squares regression analysis).[6]

To analyze the reciprocal nature of vote choice within households, we again restrict our analysis to persons who are (1) children aged 16–29 of mothers and fathers in the sample, (2) mothers/wives who live with their partner and have at least one child aged 16–29 in the sample, *and* (3) fathers/husbands who live with their partner and have at least one child aged 16–29 in the sample. The respondents are members of the same families, forming mother-father-child triads. To be included in our analysis, each of these persons must also answer the vote question in one of the election years *and* must *also* answer the partisanship questions in the two years prior to that election. As before, this focuses our analysis on two periods, 1995–7 and 1999–2001. These decisions produce a sample of more than 600 person-years for each model, more than 1,929 observations, three times the base number for mother, father, and child.

---

6 See the discussion in Chapter 5, which justifies the selection of this model. Here too we check our analysis by running a series of models using instrumental variable probit analysis. Again, the results substantially mimic those displayed in Tables 6.3 and 6.4. Applying the same test that we use in Chapter 5, we took the parameter estimates obtained from OLS and logit models and use Stata/SE 8.2 to predict linear estimates for each. They are virtually identical (results available from the authors).

Turnout and Vote Choice in Recent British Elections

We model the voting decision of each person in the family as the result of three sets of variables: (1) his or her own personal characteristics, namely partisan constancy during the two years prior to the vote, subjective social class, and whether or not the person belongs to a union (both in the election year); (2) instrumental variables to tap the likely electoral choices of the other two in the family (political interest, subjective social class and union membership, but not religious identification, because of its relatively high polychoric correlation among the members of the family), and (3) measures of aggregate political trends (the strength of each party in the region and nation, in the year of each election). We also include an interactive variable to assess whether or not the effect of others in the household is multiplicative.

The results confirm the analytical power of partisan constancy and household effects. Because our analyses find no differences between the two elections, the tables present a summary of the determinants of vote choice in 1997 and 2001. At various stages of this research, we introduced an indicator of closeness (whether each partner in the household chooses the other as the "best friend") and an interactive measure that includes the level of political interest of each member of each dyad to determine whether the more interested are more likely to influence others, to learn from the others, or to display no particular effect; neither affects vote choice. Because a party's national strength has no direct influence on any of these relationships, we exclude this variable from the presentation of the results.

Table 6.3 shows the reciprocal impact of wives/mothers, husbands/fathers, and children on each other's decisions to support Labour, after controlling for the effect of each person's partisan constancy, subjective social class, and union membership and Labour's strength in the region of residence. The results are easily summarized: for each member of the household, the frequency with which a person supports a party during the two years preceding the election always influences vote choice. No matter the power of partisan constancy, however, others in the household also influence the probability of voting Labour. Here, the impact varies by dyad: wives/mothers influence their husbands and children; husbands/fathers influence their wives, but not their children; and children affect their mothers but not their fathers. For each person, these relationships are more important than their own union membership and subjective social class and the party's regional strength (and its national strength, not shown in the table). In addition, for the children, parental effect is multiplicative not just additive. When both parents vote Labour, the young person is even more likely to join them, no matter the impact of all the other variables in the model. Note as well how much of the variance in voting for Labour is accounted for by these models. For

Table 6.3. *Family Effects on Labour Vote for Mothers, Fathers, and Children – Three-Stage Regression Model*

	Mother Votes Labour				Father Votes Labour				Child Votes Labour			
Dependent Variable	Coefficient	S.E.	Z-score	Significance	Coefficient	S.E.	Z-score	Significance	Coefficient	S.E.	Z-score	Significance
Mother votes Labour					0.31	.05	5.8	.000	0.26	.08	3.4	.001
Father votes Labour	0.26	.05	5.0	.000					0.1	.07	1.3	.182
Child votes Labour	0.27	.08	3.3	.001	−0.03	.09	−0.28	.781				
Mother Labour choice past 2 years	0.24	.02	12.3	.000								
Father Labour choice past 2 years					0.30	.02	15.2	.000				
Child Labour choice past 2 years									0.18	.02	8.1	.000
Labour support in region (%)	0.04	.14	0.3	.777	0.25	.14	1.8	.072	0.12	.17	0.7	.478
Mother union membership	0.03	.03	1.1	.255								
Father union membership					0.01	.03	0.18	.854				
Child union membership									0.01	.05	0.2	.858
Mother subjective social class	0.02	.01	−1.4	.155								
Father subjective social class					0.00	.01	0.02	.984				
Child subjective social class									−0.02	.02	−1.3	.213
Interact Mother*Father Labour vote									0.15	.08	1.9	.056
Interact Father*Child Labour vote	0.04	.08	0.5	.645								
Interact Mother*Child Labour vote					0.14	.09	1.5	.141				
Constant	0.05	.05	0.9	.346	−0.04	.05	−0.84	.401	0.01	.06	0.1	.945
n	643				643				643			
R^2	.57				.63				.32			
Chi^2	813, $p = .000$				1,031, $p = .000$				279, $p = .000$			

Endogenous variables: mother votes Labour, father votes Labour, child votes Labour
Instrumental variables: union membership, subjective social class (for mother, father, and child)

Turnout and Vote Choice in Recent British Elections

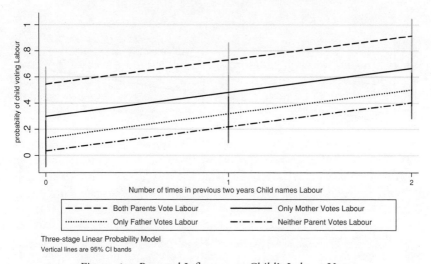

Figure 6.1. Parental Influence on Child's Labour Vote

both, wives/mothers and husbands/fathers, partisan constancy and the effect of others in the household (especially their spouse/partner) accounts for considerably more than half the variance (0.58–0.63) in the probability that each votes Labour.

Presenting the results of postestimation analyses depicts these relationships. We offer three graphs. Figure 6.1 shows the joint impact of partisan constancy and parents' vote for Labour on the probability that the child votes Labour. Note that the expected probability of a Labour vote for a child, setting each explanatory variable at its mean, is 0.32. Two parental ballots for Labour always raise the probability of voting Labour, no matter the partisan constancy of the young person. Consider the extremes: where the child always prefers Labour and both the mother and father vote Labour, the probability that he or she votes Labour exceeds 0.80, more than twice the mean. Where the young person never names Labour in the preceding years and neither parent votes Labour, the probability of Labour voting approaches zero! The difference in the probability of voting Labour between these two extremes is 80 times. Note as well that mothers have a stronger impact on this outcome than do fathers.

The next two figures display the impact of partisan constancy and the voting decisions of household partners and children on the probability that the wife/mother votes Labour (Figure 6.2) and the husband/father votes Labour (Figure 6.3). In both cases, partisan constancy has a strong impact on vote choice, and in both cases the partner's vote for Labour strongly affects the probability of casting a ballot for Labour. Note the mean levels for each partner: 0.47 for the wife/mother and 0.46 for the

Partisan Families

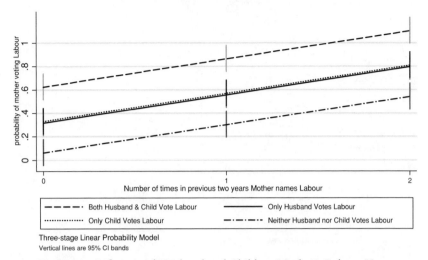

Figure 6.2. Influence of Husband and Child on Mother's Labour Vote

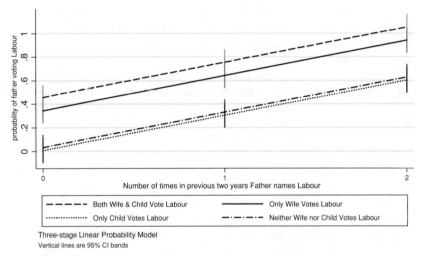

Figure 6.3. Influence of Wife and Child on Father's Labour Vote

husband/father. At the extremes, where each partner supports Labour in each of the preceding two years and the spouse votes for the party, the respondent is almost certain to vote that way, again at twice the mean level for each. Where none of these characteristics applies, the probability of a Labour vote is virtually zero. These figures also highlight the relative importance of spouse/partner compared to the child as a determinant of voting.

Table 6.4 presents the results of the models for the probability of voting for the Conservatives, for each of the three members of British households. Again, mothers/wives stand at the center of political influence within the family because they affect the voting decisions of their children and husbands/partners. Here, husbands/fathers affect their wives, but not their children, and the youngsters have no influence on their parents. Of distinctive importance here, the effect of husbands and wives on each other is greater than is their own partisan constancy. In this case, it is as if their embrace keeps them from succumbing to the force of Labour's national political hurricane. For children, however, the effect of their mothers or the combined force of their mothers and fathers is less important than their own partisan constancy as a predictor of a Conservative vote. Unlike their parents, children seem to be affected by Labour's national successes. Also, the effect of others in the household on choice is multiplicative as well as additive. For household partners, casting a ballot for the Tories as for Labour rests heavily on partisan consistency and spouse's vote (accounting for 0.57 of the variation in each partner's vote). This intimate social process helps to account for Tory voting.

Figure 6.4 displays the postestimation probability of a child's voting for the Conservatives, where the expected probability of a Tory vote, setting all explanatory variables at their means, is 0.13. While the general pattern repeats the results for Labour, the figures clearly display how much more important are mothers than fathers on the child's probability of voting Tory.

Figures 6.5 and 6.6 display the relative power of partisan constancy and the voting decisions of spouses/partners and children on the probability of British wives/mothers and husbands/fathers casting a ballot for the Conservatives. As we show with regard to Labor vote, spouses are more important than children, but the effect of children on their mothers is greater than on their fathers.

Consider now all the figures together and note the difference between the probability of voting and not voting for each party among the members of the families. Where the respondent – whether child, wife/mother, or husband/father – does not name the party in each of the two years prior to the ballot and where the other two in the family do not vote for the party, the probability of voting for the party is virtually zero. Wives/mothers and husbands/fathers whose partisan constancy is at the maximum and who live with others who vote for the party are certain to do so as well. Children, in turn, even with the same partisan history and where both parents vote for the party, approach but do not reach certainty.

Preference constancy and the voting choices of others in the household strongly and persistently influence electoral decisions. The more likely is the woman of the house to vote for a party, the more likely is her partner

Table 6.4. *Family Effects on Conservative Vote for Mothers, Fathers, and Children – Three-Stage Regression Model*

	Dependent Variable											
	Mother Votes Conservative				Father Votes Conservative				Child Votes Conservative			
	Coefficient	S.E.	Z-Score	Significance	Coefficient	S.E.	Z-Score	Significance	Coefficient	S.E.	Z-Score	Significance
Mother votes Conservative					0.59	0.05	12.6	.000	0.25	0.07	3.7	.000
Father votes Conservative	0.51	0.04	11.7	.000					−0.02	0.07	−0.3	.751
Child votes Conservative	0.11	0.08	1.4	.153	−0.06	0.1	−0.6	.524				
Mother Conservative choice past 2 years	0.19	0.02	9.6	.000								
Father Conservative choice past 2 years					0.19	0.02	11.0	.000				
Child Conservative choice past 2 years									0.22	0.02	10.0	.000
Conservative support in region (%)	−0.12	0.18	−0.7	.497	0.2	0.18	1.2	.249	0.43	0.18	2.5	.014
Mother union membership	−0.01	0.02	−0.6	.569								
Father union membership					−0.04	0.02	−1.9	.061				
Child union membership									0.02	0.03	0.6	.572
Mother subjective social class	0.01	0.01	0.6	.545								
Father subjective social class					0.01	0.01	1.0	.321				
Child subjective social class									0.01	0.01	1.3	.196
Interact Mother*Father Conservative vote	0.23	0.08	2.7	.008								
Interact Father*Child Conservative vote					0.17	0.11	1.7	.100	0.2	0.07	2.7	.008
Interact Mother*Child Conservative vote	0.04	0.03	1.2	.251								
Constant	0.04	0.03	1.2	.251	−0.01	0.03	−0.4	.667	−0.07	0.03	−2.2	.029
n	663				663				663			
R^2	.57				.57				.37			
Chi²	1,035, $p = .000$				1,047, $p = .000$				411, $p = .000$			

Endogenous variables: mother votes Conservative, father votes Conservative, child votes Conservative
Instrumental variables: union membership, subjective social class (for mother, father, and child)

Turnout and Vote Choice in Recent British Elections

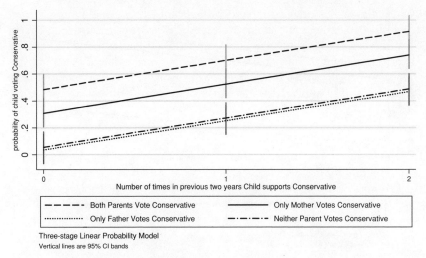

Figure 6.4. Parental Influence on Child's Conservative Vote

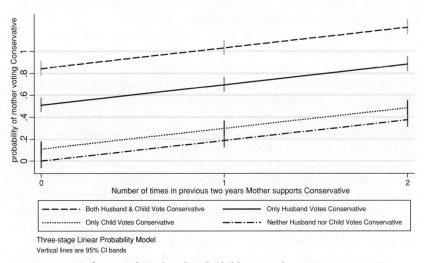

Figure 6.5. Influence of Husband and Child on Mother's Conservative Vote

and child to do so as well. The more likely the man of the house is to vote for a party, the more likely his wife is to vote for the same party, and the more likely the child is to vote for a party, the more likely his or her mother is to vote for the same party also, regarding Labour. The greater the level of partisan constancy is, the more likely people are to vote for the party. Recall that in Chapter 3 we show that preference (or partisan) constancy

Partisan Families

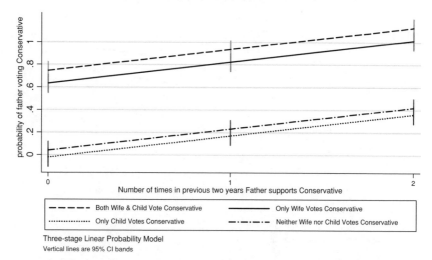

Figure 6.6. Influence of Wife and Child on Father's Conservative Vote

responds to the political preferences of social intimates. Together, these results sustain the social logic of politics.

Again, women stand at the political center of the social relationships that are households or families. They always influence the electoral choices of their children and their husbands; husbands always affect their wives' ballots, and children are more likely to influence their mothers than their fathers. The results of the linear probability models with instrumental variables show that with regard to politics (and perhaps many other matters), in families social influence flows through the senior woman.

Consider now how household members in combination affect each other. As we saw in the previous chapter with regard to partisan preference, the probability of choosing Party A is always higher where two in the household pick it, than when one does. This is certainly the case for wives/mothers and children, but for husbands/fathers, the joint effect of the others is relatively weak. Note too that in order for the children to influence the husband/father's choice of Labour, the wife/mother must also vote for Labour. Again, the combined effect of two members of the household adds to the probability of not voting for Party A.

The particular results of our analyses point to the importance of frequent interactions, past learning, authority, shared identity, and content specificity as the bases of cue-giving and taking in social relationships. We find no evidence to support the claim that political interest by one or the other or both members of a dyad influences political learning in the household. As well, there is no support for the hypothesis that perceptions of social closeness influence this process.

Turnout and Vote Choice in Recent British Elections

Having established the importance of intimate social relations on partisanship and voting, we move our analysis beyond the microworld of households and families to link our argument to broader issues of political parties and campaigns. In the next chapter, we show how election campaigns reinforce processes that occur in households, establishing electorates that are composed of bounded partisans. Before we turn to that analysis, we present evidence that sustains our emphasis on the power of households and families on political decisions.

A COMPLEMENTARY ANALYSIS OF THE SOCIAL LOGIC OF TURNOUT IN BRITAIN

Drawing our data from two exceptionally useful surveys of Germans and Britons over time, we present robust findings that support the social logic of partisanship. No matter the strengths of GSOEP and BHPS – and they are exceptional – they also have weaknesses. The surveys provide snapshots of each member of the household, at one point in time and averaged over time as well as over multiple points in time, but always one member at a time. Because they offer few questions that directly depict relationships within households (and none of these examine politics), our analysis draws inferences about similarities and reciprocal effects from statistical manipulations – from cross-tabulations to linear probability models. Now, we extend our analysis by examining survey results that include measures of conversations about politics, a more direct assessment of the politicization of social ties.

The Local Participation Study in the United Kingdom[7] includes questions on the frequency with which the respondents discuss politics with spouse or partner, parents, children, workmates, neighbors, and friends. It also includes other social variables directly associated with the decision to go to the polls: whether or not they live with another adult and the amount of time that they devote to attending meetings of organizations and to interacting with and helping other persons. Table 6.5 presents a logit model that examines whether or not the respondents have voted in a general election. Table 6.6 applies an ordered logit model on the frequency of voting in local elections ("never," "sometimes," and "always" are the options).

Again, household effects appear. The frequency of political conversations with spouse or partner has a statistically significant effect on both

---

[7] Colin Rallings and Michael Thrasher direct the study; it was administered in 2002; UKDA #4849. We report the data from 2001. The analysis is our responsibility, and we thank the directors and the U.K. Data Archive for making them available.

Partisan Families

Table 6.5. *Participation in British General Elections – Logit Model*

	Coefficient	S.E.	Z-Score	Significance
Party support	1.42	.12	11.6	.000
Interest in local politics	0.33	.09	3.7	.000
Frequency discusses politics with spouse	0.11	.02	4.4	.000
Amount of time spent in organizations	0.09	.03	2.6	.010
Length of time living at residence	0.11	.07	1.6	.105
Feel guilty if do not vote	0.35	.05	6.8	.000
Age	0.13	.04	3.1	.002
Constant	−2.54	.29	−8.8	.000

Data source: U.K. Local Participation Study
$n = 2,819$
Log likelihood = −1001
$Chi^2 = 472, p = .000$

Table 6.6. *Frequency of Participation in British Local Elections – Ordered Logit Model*

	Coefficient	S.E.	Z-Score	Significance
Party support	1.18	.10	11.5	.000
Frequency discusses politics with spouse	0.13	.02	7.3	.000
Amount of time spent in organizations	0.16	.03	6.1	.000
Length of time living at residence	0.18	.05	3.5	.000
Feel guilty if do not vote	0.61	.04	15.8	.000
Age	0.36	.03	11.2	.000
Cut 1	2.57	.21		
Cut 2	4.95	.23		

Data source: U.K. Local Participation Study
$n = 3,004$
Log likelihood = −1,926
$Chi^2 = 1,053, p = .000$

forms of turnout, when controlling for obviously important predictor variables: partisan support; interest in local politics; the extent to which the respondent would feel guilty about not voting; age; education; the length of time that they lived at their address; and the strength of their attachment to their neighborhood or village. Furthermore, the other measures of political conversation have no impact on turnout, net of the other predictors. In this model, living with others in and of itself has no effect on turnout, implying that propinquity is not the causal agent. Political conversations that move along the intimate ties of marriage and household partnership help to determine the decision to go to the polls.

CONCLUSIONS AND IMPLICATIONS

The micropolitics of families helps to structure the national electorate. Persons who never choose Party A/B in the two years prior to an election and who live with others who do not vote for the party cast their ballots for another party or stay home. Wives/mothers and husbands/fathers who display preference constancy and live with others who vote for Party A/B do so as well, so do their children who have high partisan constancy, but at a slightly lower probability.

Viewed from the perspective of the British electorate taken as a whole, most people (0.65) live with others and most (0.68) of them receive consistent voting cues from others in their households and families. About one third of those who cast a ballot for Party A/B live with one or two persons who also vote for Party A/B; half live and/or have intimate social ties with two persons neither of whom supports Party A/B; 0.07 live with two persons who do not vote; and 0.10 live with persons all of whom cast a ballot for one of the other parties, almost always the Liberal Democrats. Very few live in households in which the others present split their votes between the major parties or someone votes for a major party and the respondent votes for its rival (0.07) or a major party and another party (0.03).[8] As a result of these processes and the other social ties that structure the preferences of those who live alone, approximately 0.80 of the British electorate enters election campaigns already committed to one of the parties (see Johnston and Pattie 2005, 186, with regard to the general election of 2001, citing British polls).

In the next chapter, we develop this final theme. There, we move from the level of the primary group to the electorate taken as a whole. We tie politics within the household to politics within the nation. Each level influences the other. The social logic of bounded partisanship guides our analysis.

---

8 Zuckerman and Kotler-Berkowitz (1998) find similar aggregate patterns in the 1992 elections, when many more households displayed electoral cohesion for the Conservatives.

Conclusion
Family Ties, Bounded Partisanship, and Party Politics in Established Democracies

We apply a social logic to mass politics. Enmeshed in intimate social ties, most people live their lives far from the world of politicians and candidates and from the print, radio, television, and internet media. No matter the efforts of the party leaders and political experts, conversations especially among family members – verbal and nonverbal, direct and indirect – refract political messages. When people consider the political parties and when they vote, they bring along the expectations, and they influence the preferences of the members of their personal networks. No matter the isolated circumstances surrounding survey responses and no matter the curtains that surround the voting booths, when people think about the political parties and when they cast ballots, they do not sit or stand by themselves.

Wives/mothers are at the heart of this process. Their partisan preferences consistently affect and are affected by the members of their families. Examining why they are central to the formation of bounded partisans in Germany and Britain directs attention to the social mechanisms that account for social influence, the bases of bounded partisanship. It also suggests the factors and conditions that structure and might transform the party systems in Germany and Britain.

In this chapter, the conclusion to our volume, we also relate the immediate social context of partisanship and electoral behavior – the family – to party politics – competition between and among the political parties. From the perspective of the individual, we link the individual's most directly experienced reality to a world perceived only through distant others – reporters, broadcasters, and various other experts whom they read, hear, or see and party activists who come knocking on their doors. Just as there are reciprocal relationships of influence within households, politics in intimate locations interacts with national politics. The social logic of politics induces the presence of bounded partisans, and their presence

Conclusion

induces campaign activists to ignore those least likely to vote for them and to tailor their appeals to their supporters. In turn, the party system influences the strategic and tactical decisions of party leaders, and the implementation of their decisions affects the partisan choices of citizens. These reinforcing processes structure bounded partisanship in Germany and Britain.

It is important to keep in mind that we examine German and British politics at a particular moment. We sketch the bases of a persistent set of interactions that occurs in the two countries during the past two decades. We do not suggest that these structures are permanent, far from it. Even a cursory understanding of the political history of Germany would caution against that assumption, and recent Italian politics reminds the observer that even the most stable party systems can collapse. And so, in this chapter, we suggest that the sources of German and British political stability contain the seeds of their own transformation.

ON THE RELATIVE POLITICAL IMPORTANCE OF WIVES AND MOTHERS IN HOUSEHOLDS

While wives and husbands influence each other's political preferences in a generally symmetrical manner, women as mothers have much stronger and more consistent impact on their offspring than do men as fathers. Our findings show that fathers and children sometimes influence each other in positive ways, sometimes in a negative manner, and sometimes not at all. Almost always positive, the reciprocal relationship between mothers and their offspring offers a much simpler picture.

Why do mothers and children have a positive reciprocal relationship with regard to partisan choice and electoral decisions? Why is the relationship between father and child inconsistent, and rarely as strong as that between mother and child? In the first chapter, we sketch an approach that accounts for the variable strength of the dyadic relationships within households that rests on the frequency of interactions and the presence of trust and recurrent learning. This implies answers to these questions: children spend more time with their mothers than with their fathers; they feel closer to their mothers than their fathers, and so they learn more from their mothers than from their fathers; and because they spend more time with their mothers, they are better able to influence them as well.

There is much evidence to support these generalizations. First and with specific reference to politics, we again draw on data from the U.K. Local Participation study (neither GSOEP nor BHPS speaks to this particular issue). Whereas British men and women who live with another adult are equally likely to speak to their spouse or partner, women have higher rates of political conversations with their parents and their children and lower rates with neighbors, workmates, and social friends.[1] Simply put, women are more likely than men to talk politics with family members.

More generally, research on how persons spend their time affirms again and again that mothers interact more frequently with their children than do fathers. Evidence on Germany and Britain details patterns that reappear across Europe, the United States, and Canada (Coltrane 2000; Eurostat 2003; Gauthier, Smeeding, and Furstenberg 2004; Joesch and Spiess 2002; Sayer, Bianchi, and Robinson 2004; Sayer, Gauthier, and Furstenberg 2004). First two general points: most people spend most of their time at home and women are more likely than men to spend 0.60 or more of the day at home (Eurostat 2003, Tables 8.1, 8.2, 8.3, 108–10). In households in which the youngest child is 7 to 17 years old, women, whether or not they are gainfully employed outside the home, spend twice as much time as husbands on "domestic" work (more than four hours for the female; Eurostat 2004, 8, Table 6.2, 78). Each day, German and British women devote on average approximately half an hour directly to childcare, while their husbands or partners spend less than twelve to thirteen minutes with the children (Eurostat 2003, Table 5.13, 68). In recent years, fathers give more time to childcare, and still mothers interact more frequently with their children. In Germany, 0.73 of the mothers and 0.16 of the fathers spend "substantial parental time" (defined as more than twenty-seven hours per week) with the children. The proportion is not much different in the United Kingdom (0.84 of the mothers and 0.23 of the fathers; see Smith 2004). Using GSOEP data, Cooke (2004) reaffirms these descriptions, detailing that West German women spend on average thirty-two hours per week on childcare, four times the rate of fathers. Drawing on related data, Fisher, McCulloch, and Gershuny (1999) report that British mothers of children ages 5–15 each day spend nearly an hour with their youngsters, and British fathers spend forty minutes.

GSOEP and BHPS shed direct light on these matters. Each provides answers to a battery of questions that were asked to a sample of adolescents, youngsters between the ages of 11 and 16. The German data are particularly extensive. The youngsters frequently (defined as "often or very often") talk with their parents about their lives, and they consistently report higher rates of these conversations with their mothers. In order to detail the level of frequency, we will highlight the responses: "Talks about

---

[1] As we cite in Chapter 4, Miller, Wickford, and Donaghue (1999) note that in Northern Ireland women are more likely than men to find their discussion partners among family members.

Conclusion

things that interest you:" mother (0.60), father (0.40); "Talks about things that worry you:" mother (0.41), father (0.28); "Asks you prior to making decisions:" mother (0.61), father (0.50); "Expresses an opinion on something you do:" mother (0.75), father (0.54); "Able to solve problems with you:" mother (0.57), father (0.46); "Appears to trust you:" mother (0.78), father (0.70); "Asks your opinion on family matters:" mother (0.50), father (0.39); "Gives reason for decision:" mother (0.52), father (0.45); and "Shows that she or he loves you:" mother (0.78), father (0.65). BHPS's adolescent sample reports only a few of these dimensions. For example, 0.57 of the British youngsters talk to their mothers and 0.35 to their fathers about "things that matter" to them "more than once a week." They are more likely to fight with their mothers than their fathers, but they also interact much more frequently with the mothers. Nearly two thirds of the British adolescents report eating at least three evening meals with their families each week, with most of these having six or seven such meals together. Most German and British adolescents are closely tied to their mothers and fathers, especially their mothers.

We have found no evidence that political influence within households follows from political interest and by extension from political knowledge. Fathers are no more able (with regard to the SPD) and less able (the other three parties) to influence their children's partisan choices, even though fathers are usually the parent who displays more interest in politics. Wives always influence husbands, no matter that they are generally less interested in politics. In Germany, the mean levels of political interest (on the scale that we use in Chapter 2) are 2.1 for children, 2.1 for mothers, and 2.6 for fathers. In Britain, the results are 2.08 for children, 2.2 for mothers, and 2.8 for fathers. Within households and perhaps elsewhere, political influence rests on the frequency of interaction and emotional bonds, not political interest.

Children learn more from their mothers, and so they also learn more about politics from them. They are more likely to teach their mothers than their fathers, and so they are more likely to influence their mother's political choices. These relationships underscore the reciprocal nature of household partisanship. In turn, they reinforce the likelihood that people are bounded partisans, whose choices taken in the aggregate move the analysis from the family to national politics.

THE SOCIAL LOGIC OF BOUNDED PARTISANSHIP

Stated generally, Germans and Britons are bounded partisans. Most never support one of the major parties and vary their selection of its major rival, moving back and forth over time between picking it and saying that they prefer no party. The adjectives "consistent" and persistent" apply best to the party not chosen, not to the party usually named. The probability of naming a party in a survey or at the ballot always varies; high preference constancy describes few Germans and Britons. Variation in the political preferences of household members influences all dimensions of partisanship and electoral behavior. These statements reappear many times in our volume.

These findings reaffirm the utility of the directional model of electorates (see, for example, Diamond 2001; Macdonald, Listhaug, and Rabinowitz 1991; Rabinowitz and Macdonald 1989) and the presence of negative partisanship (Crewe 1980; Särlvick and Crewe 1983). As important, they again question Downs' decision to ignore the effects of household members on each other's partisanship and electoral choice (see the discussion in Chapter 1). This analytical choice induces scholars to posit the presence of a median voter, rather than the bounded partisan, who stays on one side of the national political divide.

Our models and simulations allow us to detail these summaries. By definition, each person taken as an "average" member of the population shares the mean probability of support for each party, Party A/B, in the population at large. Reciprocal influences within households and families place, lower, and raise the probability of support with regard to the population mean. Drawing on the Heckman Probit selection models and the attached simulated predicted probabilities that we present in Chapter 3, we show that where the net balance of partisan preferences in the head of household's family is zero, he or she stands at the sample mean of preference for A/B. Where the net balance is negative, he or she is certain not to name Party A/B. Where it is positive, the probability of preference rises well above the mean.

The linear probability models and the accompanying simulated predicted probabilities in Chapter 5 allow us to specify these relationships among children, husbands/fathers, and wives/mothers. Consider first the results in comparison to the mean level of partisan preference for each. For children and wives/mothers in households in which neither one of the others names Party A/B, the probability that each picks the party ranges from 0.40 of the mean (women for the Conservative Party) to slightly more than 0.01 (children for the Conservatives). The effect on husbands/fathers is not much weaker. In the absence of reinforcing cues and messages of others in the intimate social circle, the rate of partisan preference drops well below the mean. Furthermore, the opposite choices of husbands and wives always place their partners above or below the mean; mothers have the same effect on their children, and children influence their mothers in concert with their fathers, but do not have a positive effect on their fathers by themselves. Household effects

Conclusion

influence the probability of partisan preference relative to the mean rate of selection.

Now consider bounded partisanship in absolute terms (i.e., not in relation to the population mean, again drawing first on the results that we present in Chapter 3). First and foremost, it is much easier to reach certainty with regard to *not* naming a party. The simulations based on the initial models indicate that a negative balance of one ensures that a person does not support Party A/B. In these analyses, a positive balance of two implies a variable level of support for Party B/A. Indeed, the postestimation simulations highlight how unusual it is for the probability of support to approach statistical certainty. Only rare combinations of social indicators induce this outcome.

The results of simulations derived from the linear probability models reinforce these findings (Tables 5.8–5.10). Where two members of the family name party A/B, mothers and children are virtually ensured not to pick Party B/A. The certainty of preference is as strong only for the selection of Labour among British husbands/fathers, whose wives and children also name Labour. As a result of this social logic, most people are bounded partisans; they never support Party A/B and vary their choice of Party B/A.

Not only do the reciprocal influences within households have an especially powerful impact on the probability that a person does *not* name a party, the distribution of preferences within households reinforces this pattern. As we note in Chapters 3 and 4, few Germans or Britons live in households in which everyone chooses the same party at the same time. Consider here only the three-person households, drawing on the results that we show in Chapter 5. In Germany, very few are characterized by everyone naming the SPD in an average year, and in only slightly more of the households does everyone pick the CDU/CSU. At the other extreme, in more than half of the households no one names either party. The patterns are similar in Britain. There, few families are composed of everyone present picking Labour, and in even fewer do all name the Tories. Again, many more families are cohesive with regard to the party not named. The immediate social context of most people's lives reinforces the decision to not name a party and varies in the extent to which it induces preference for its rival.

The social logic of bounded partisanship implies that people apply cues from family members to construct a choice set from the full list of political parties. They eliminate one of the major parties and establish the other as a possible object of support. Messages from family members help to account for the probability of partisan choice at any point in time and over time. Variation in preference constancy combines with the decisions of family members to help explain electoral behavior.

Partisan Families

A RECIPROCAL RELATIONSHIP BETWEEN BOUNDED PARTISANSHIP AND PARTY POLITICS

Party leaders in contrast to citizens, make strategic decisions in a contest with a clear goal and defined rules. They seek to turn the assent of voters into political action (ballots at the polls, funds, and voluntary labor). Suggesting that these political actors strive to amass the optimal number of votes (the maximum needed to win divided by the costs of the effort) makes perfect sense. It provides their "macroscopic driving force" (Aumann 2005). This applies whether winning means conquering the heights of power or maintaining a foothold in the legislature. The persons – citizens, members of the public – whose responses we analyze in this volume exist in a different world. No clear goal focuses their perceptions of the political parties. No strategic choices follow. Even if the principles associated with rational choice theory may not help us to understand most voters, they allow us to interpret most party leaders (Meehl 1977 and Riker 1982 provide the classic basis for this distinction).

In Chapter 1, we show that Key and Downs lead a transformation that blurs the distinction between how citizens and politicians perceive partisanship. Emphasizing the primary group, they insist, takes this study from the realm of politics to the world of small groups. Our research rejects the need to choose between a theoretical perspective that emphasizes the family and one that focuses on politics. Rather, there is a reciprocal relationship between these two spheres of life (and thus we sustain the observation with which the behavioral revolution begins; see Chapter 1).

No matter the fundamental differences that separate party leaders and citizens, they are tied by the effects of bounded partisanship. Seeking to optimize the balance between the costs of electoral campaigns and the maximization of votes, party leaders construct an object set from the full electorate. They divide voters into categories or groups, according to the likelihood of vote for the leaders' party at the polls. As they proceed, they consider how bounded partisans are aggregated into national political divisions. In turn, the actual partisan divisions that follow from the particular choices of bounded partisans affect how the leaders construct the object sets among the voters.

For some party leaders, winning implies complete victory; for others, it suggests a shared place in the governing coalition or seats in the legislature. In Germany, the relatively equal size of the Social Democrats and Christian Democrats/Socials and the balance between their preferred coalition partners induce the leaders of the SPD and CDU/CSU to seek to control the government by wining more votes than their main rival. A plurality guarantees control over the governing coalition. In Britain, electoral rules usually allow the party that garners the most elected representatives

Conclusion

to control a majority in the House of Commons and to monopolize the cabinet. In practice, this means that victory is possible only for Labour or the Conservatives, not the Liberal Democrats, or any of the small parties. The leaders of the major parties in Germany and Britain strive to win by attracting a plurality of the vote; they do not strive just to sit in the Bundestag or the House of Commons.

In Germany and Britain, the presence of bounded partisans structures partisan divisions and the choices of party leaders. Summarizing our analyses, we find that, at any point in time, four sets of persons compose the electorate: (1) those who never prefer Party A and never name party B, (2) those who never choose Party A and sometimes select Party B, (3) those who never name Party B and sometimes pick Party A, and (4) a mixed group, of which the smallest set contains people who move between the two major parties. Recall too that alignment with a party contains a relatively small number of persons who are certain to name the party and a much larger group who are more likely than not to choose the party. These distributions influence the strategies of party leaders.

The national political divisions that follow from the presence of bounded partisans imply that no party enters an election with the guaranteed support of a majority of the electorate and that their core is divided into sure and unsure supporters. They suggest as well that the probability of votes from persons who are outside the party's core constituency varies between zero and very low. Party leaders know that it is difficult to impossible to attract those who never voted, for the party especially those who belong to their rival's core constituency. Furthermore, the politicians know that the more that they appeal to outside groups, the more they risk turning away their probable but unsure supporters. To underline the critical but obvious point: for party leaders, elections are contests with clear goals and rules but unpredictable outcomes.

As campaigns unfold, party leaders broadcast and tailor their appeals to the citizenry. They use the media and public speeches to reach all voters and to make segmented appeals to particular voters. Bounded partisans usually pay little attention to the other party's appeals. Low levels of political interest and selective attention suggest that they pay some attention to their own party's efforts. In principle, door-to-door campaigns bring the political parties to each voter. In practice, manpower costs induce party leaders to create a choice set, separating those who are likely to vote for them from those who are certain not to do so. As parties recognize the high costs of trying to reach beyond their core and as they rely on the established social ties of their campaign activists, they direct their strongest efforts back to their known supporters. Assuming the support of certain voters, they direct particular attention to persons who voted for them in the past but did so in an inconsistent manner. By design and as a

Partisan Families

consequence of variations in success of mobilization, electoral campaigns reinforce the bounded nature of partisan choice.

Beginning with Berelson, Lazarsfeld, and McPhee (1954), numerous studies of campaigns support these claims. Holbrook and McClurg (2005) review the vast literature on American election campaigns, noting the distinction between mobilization and persuasion and demonstrating the differential impact of campaigns on the turnout of partisans (and see also Wielhouwer 2003). Analyzing the German elections of 1990, Finkel and Schrott (1995, and see also Finkel 1993 on American election campaigns) show that campaigns generally reinforce the predispositions of voters, noting that 0.14 of the electorate altered their previous vote. In the conclusion to their edited volume on campaigns in many democracies, Schmitt-Beck and Farrell summarize the relationship between campaigns and voters: "The more they are reached by campaigns, the more those citizens with ideological or partisan affiliations tend to support their own elites." Conversion is possible, but unusual (2002, 186–7, and see also Bergmann and Wickert 1999 on German elections in the 1990s).[2] In an analysis of German elections in the 1990s, Bergmann and Wickert (1999) caution against claiming that campaigns persuade voters to change their preferences. Applied to Britain, Bannon (2003) offers a wide-ranging analysis of the parallels between marketing efforts by businesses to attract consumers and those of political parties to attract voters. He suggests that in worlds of imperfect competition, decision makers segment their markets and focus on target groups. Similarly, Plassner, Scheucher, and Senft's (1999) study of campaign professionals in Europe notes that "target groups" are the most frequent focus of campaigns.

Scholars have devoted particular attention to constituency contests in Britain. There, party activists devote primary attention to finding persons who have voted for the party in the recent past, to ensuring that they do so again and to making sure that they get to the polls. There is little evidence that they devote any but minimal attention to known supporters of their opponents. Challengers are more likely, however, to reach beyond their core constituency than are those who work on behalf of the governing party (see Bannon 2003; Denver and Hands 2002; Denver, Hands, and Henig 1998; Denver et al. 2003; Denver, Hands, and MacAllister 2004; Holt and Turner 1968; Whiteley and Seyd 1998; 2003). Seeking people who will vote for their candidates, party campaigns reinforce the presence of bounded partisans.

---

2 For a popular analysis of how Karl Rove, chief Republican strategist in 2000 and 2004, has introduced new forms of targeting in American election campaigns, see Lemann (2003).

Conclusion

Most broadly put, election campaigns in Germany and Britain emphasize direct competition between the major parties. In turn, they further reinforce the tendency for persons to perceive partisanship as a limited choice, "picking a side, by avoiding a side."[3] This reinforces the understanding that partisanship is about locating oneself relative to the two "sides."

The result is a reciprocal relationship between the social logic of partisanship and the behavioral effects of the strategic calculations of party leaders. Because campaigns seek to mobilize persons who are likely to support Party A/B, they reach persons whose intimate social ties induce them to support Party A/B. These people do not come into contact with many supporters of Party B/A, and that party does not directly approach them. The strategic choices of party leaders reinforce the bounded party system that each person constructs and the likelihood that each person remains a bounded partisan. As a result of both reciprocal processes, Germans and Britons are highly likely not to choose Party B/A, and they vary the probability of naming Party A/B. The presence of bounded partisans in turn structures the strategic choice of party leaders.

Consider the implications of this argument for the model of turnout and electoral choice that we present in Chapter 6. Families are not isolated political units (a point noted at the start of the behavioral revolution in political science, but soon set aside; recall the discussion in Chapter 1). To the contrary, one or more of the members of the family (usually the husband/father who spends more time outside the household) brings home information, impressions, and evaluations of the candidates. In this process, mothers, refracting these cues, matter more than fathers for the children; husbands and wives affect each other; children usually influence their mothers but not their fathers. The centrality of family ties also suggests that these discussions stand apart from and influence each person's political exchanges with more knowledgeable members of their social networks. The relations depicted in Chapter 6 highlight the paths by which the world of national elections reaches households. As a result, our model offers the proximate causes of turnout and electoral choice. A full model would bring the communication flows of the campaign into the households and model the interactions between the intimate and the national levels of politics. Alas, we do not have the data (nor do we know of the existence of such data) to model this multilevel reciprocal relationship.

---

3 Duverger (1967) offers the classic source for a dualist conception of party competition, and see Sniderman (2000) and Sniderman and Bullock (2004) for parallel applications that focus on the organization of the survey instrument.

Partisan Families

In the next section, we continue to explore the social logic of partisanship by examining new entrants to the German polity, persons born in East Germany and immigrants. Again, we depict bounded partisans.

BOUNDED PARTISANS AMONG EAST GERMANS AND IMMIGRANTS

During the seventeen years of GSOEP survey data that we examine, East Germans and immigrants enter an established democracy. They have distinctive characteristics, which might be expected to affect their partisanship. Residents of the former German Democratic Republic (GDR) are new citizens of the Federal Republic; they have little experience with its political parties and no history of democratic rule. Most immigrants, on the other hand, are not citizens. Although those who come from member states of the European Union may vote in local and European elections, they do not cast ballots in German national politics. Because they bring different personal and political histories from their countries of origin, one would expect to see these differences influence their partisan choices. In addition, because there is a flow of immigrants into the country over time, these persons vary in their exposure to German politics. Even as partisan choice in each subpopulation reflects factors unique to that group (for example, views of the German Democratic Republic among East Germans, and familiarity with German life among immigrants), the members of each of the subpopulations respond to variables associated with the social logic of partisanship.

Zuckerman and Kroh (2005) analyze partisanship among those who respond to GSOEP questions in every year between 1990 and 2003. Here too they find bounded partisans. Most immigrants (like West Germans) name one of the major parties at least once and then vary their selections of that party with no announced preference, not with the other major party. East Germans create their choice set from the two major parties and the Party of Democratic Socialism (PDS), and so they eliminate two of the parties and vary their decisions between their party and no party. How do East Germans and immigrants differ from the veterans of an established democracy? West Germans display the highest levels of partisan support, and the immigrants, the lowest levels. Political influence within households appears as well among all three groups. No matter the importance of attitudes towards the former East German regime and other variables as predictors of partisan support and choice among East Germans and no matter the importance of social integration in Germany and other predictors among immigrants, the balance of partisan support and preference in the households has a strong influence on partisan support and preference.

Conclusion

Entering an established party system, East Germans and immigrants construct a set of choices from the list of political parties. The broad social contexts of their lives – the more distant elements that refract social class, religion, and language – join with the exchange of political cues within households to reinforce a pattern of bounded partisanship.

THE FUTURE OF BOUNDED PARTISANSHIP IN GERMANY AND BRITAIN

Examining nearly two decades of German politics and more than ten years of partisan behavior in Britain, we depict a structured process that rests on household routines. Furthermore, we suggest, strategic decisions of party leaders interact with this process. Each reinforces the other. These patterns withstand the exogenous shocks that accompanied Reunification in Germany, the replacement of the mark by the euro, and persistent high rates of unemployment. In Britain, they continue despite Labour's overwhelming victory in 1997, the decline of the Tories, and a massive and growing privatization of the economy. In both countries, they remain, even as the politics surrounding immigrants rises in prominence. They continue even in the face of a secular trend of decline in partisan support. No matter these social, economic, and political changes, the social logic of bounded partisanship persists.

Our analysis points to two potential sources that would accelerate the declines in partisanship and prepare the basis for a fundamental alteration of the party systems. Again, we emphasize the interaction between processes in households – particularly the role of women – and the strategic choice of party leaders.

Consider first the role of party campaigns on the persistence of bounded partisanship. Optimizing campaign resources, party leaders pay particular attention to those voters who are likely to support them. Face-to-face contacts with a constructed set of voters strengthen the wills of those who are likely to back them, and fleets of cars bring these people to the polls. Consider now the implications of decreasing levels of partisan preference on the relative balance between possible but uncertain voters and completely uncertain voters. Recall that many Germans and Britons offer no partisan preference. Indeed, there is evidence that in the German elections of 2005, as many as 0.40 of the voters made their electoral decision in the last week of the campaign.[4] If and when the party leaders

---

[4] As reported by Rüdiger Schmitt-Beck in an email message to Zuckerman; an observation based on the preliminary results of a survey of the 2005 German elections. Holzhacker (1999) observes increasing numbers of late deciders, and Schmitt-Beck (2006) offers the general observation that increasing numbers of voters are making

calculate that there are too few core voters (the certain and the probable but uncertain) to bring them close to victory, they will consider changing their campaign strategy and tactics. They will choose between targeted (or segmented) and broad campaigns. The former will continue bounded partisanship, reinforcing old and finding new social contexts that structure vote choice. The latter will erode bounded partisanship.

Consider next the fundamental role of wives/mothers in the persistence of bounded partisanship. Now recall the declines and relatively low levels of partisan support and preference especially among women. If this trend continues, a process by which wives reinforce their husbands' preferences and teach their children the basics of partisan choice will reach a tip-over point and head in the other direction. Then, wives and mothers will dampen the level of partisan support among their husbands and not raise the level among their children. As time passes, these alterations will accelerate. Changes among the wives/mothers will have a multiplier effect, increasing the rate of partisan decline across the electorate, inducing the party leaders to reach beyond their shrinking core constituencies. A spiral in which the parties increasingly emphasize broad appeals follows, further eroding bounded partisanship. Again, party leaders will choose between further broadening the source of voters and segmenting the political market. The former will end bounded partisanship, and the latter will reinforce it.

Our evidence suggests that Germany, where more voters are uncertain at election time and where more wives/mothers choose no party, is closer to the tipping-point than Britain.[5] Until one or both of these happens, processes central to the family and party politics will continue to reinforce the presence of bounded partisanship in Germany and Britain. This reciprocal relationship structures partisan politics in these established democracies.

We have explored the details of the social logic of bounded partisanship in Germany and Britain. Our data provide a means to peer directly into households. Using these data, we have described and analyzed the place of politics in intimate social relations in two societies over a limited period

---

electoral decisions during the campaign; Johnston and Pattie (2005) cite a lower level of available voters in Britain (see Chapter 6).

[5] Applying a very different theoretical perspective to GSOEP data, Edlund, Pander, and Haider (2005) offer another source of change: the association between declines in the rate of marriage among young women and their increasing preference for left-wing political parties. Here, women are central to the prospects of policy change. For a broad overview of the relationship between variations in government policy in three German regimes – the Nazis, the Communists, and the Federal Republic – and the marital status of women, see Heineman (1999), and for a focused comparison of East and West Germany, see Cooke (2004).

Conclusion

of time. We use GSOEP's and BHPS's fine-grained data to address broad goals of analyses. Our theoretical ambitions induce us to push our models beyond our data. Even so, we recognize that only data like the ones that we have examined would permit the same kind of detailed analyses. Absent household panel surveys, we can reason and speculate, but we cannot test our hypotheses.

THE SOCIAL LOGIC OF PARTISANSHIP IN OTHER ESTABLISHED DEMOCRACIES

There is every reason to expect people to create choice sets in other party systems – to be bounded partisans. The first and easiest generalization would apply to other political systems with two dominant parties, like the United States. The expectation would also apply to party systems with a clear left-right division, such as those found in the countries of Scandinavia. In the case of one large party on side A of the political spectrum and several relatively equal parties on side B, we would expect persons who locate themselves on side A (1) to reject the other set, never supporting any of them, and (2) to vary their selections between their party and no preference. Another large group of citizens will move among the choices on side B (including no party), without ever choosing the large party on the other side of the political fence. Where there are many relatively small parties, we would expect people to frame their choices into two blocs and then to vary their support among the parties located in one of the two divisions. Where the parties are overturned, as has happened in Italy, we would also expect citizens and politicians to construct choice sets together with members of their households. Drawing on data from the Carlo Cattaneo Institute, we have examined the results of Italian election surveys over the past decade. Here, the Christian Democrats, Socialists, and Communists, the political parties that dominated politics for more than four decades, collapsed or transformed themselves into new political parties, and *Forza Italia*, along with the other new parties, and a revamped movement with fascist roots entered the party system. No matter the several dozen parties that stand for election, the voters and politicians have moved towards the creation of two competing blocs. Voters seem to choose among the left/right set of parties, never voting for those on the right/left. Politicians frame contests around the two poles, presenting candidates of the left and right in debates against each other and insisting that the winning set of political parties would monopolize the governing coalition. Here too there is evidence of bounded partisanship.

In turn, this implies that students of political behavior should incorporate the effects of personal networks in their models of partisan and electoral choice. They should also recognize and adjust their thinking to

Partisan Families

the presence of reciprocal relations among variables. At a technical level, absent these variables, the models are misspecified. More broadly, including these variables and relationships brings analysis much closer to how people perceive and make political decisions. As an aid to the reader, we now summarize our theory, hypotheses, and empirical results.

IN SUM

Social interactions lie at the heart of people's lives. These reciprocal relationships, we emphasize, are not closed and not determined. As a result, all outcomes and relationships are probabilities. Our analyses place each person in relation to the mean response and to the absolute standards across population averages.

Concepts and Their Measures

We examine several elements of partisanship: partisan support (whether or not a person names a party), partisan preference or choice (the party named), and partisan constancy (the rate of selection over time). Turnout (going to the polls) and vote choice (the party or candidate named in the ballot) compose electoral behavior. Each concept is distinct from the others. Analytically, partisan support is a precondition of partisan choice, and turnout precedes voting. There is an empirical relationship between preference constancy and electoral choice.

Almost all of our empirical evidence comes from the German Socio-Economic Panel Study (for the years 1985–2001) and the British Household Panel Survey (1991–2001). Looking at a few other surveys, we perceive households and politics through the responses to GSOEP's and BHPS's questions.

GSOEP defines the elements of partisanship with a set of related questions. In English, they read: "Many people in the Federal Republic of Germany [Germany, after 1990] are inclined to a certain political party, although from time to time they vote for another political party. What about you: Are you inclined – generally speaking – to a particular party?" We define those who respond "Yes" as party supporters. They are then asked, "Which one?" without a list of the parties. This answer taps partisan choice. As in the German data, repeat responses affect support and preference constancy.

BHPS offers three questions: "Generally speaking do you think of yourself as a supporter of any one political party?" If the answer is "No," the survey then asks, "Do you think of yourself as a little closer to one political party than to the others?" We define those who say "Yes" to either question as party supporters. They are then asked, "Which one?" without

Conclusion

a list of the parties. This answer taps partisan choice. As in the German data, repeat responses affect support and preference constancy.

We focus on partisanship among people who live together: heads of household in relation to others who live with them, spouses/partners (men and women who may be married or not), and parents and children (between the ages of 16 and 29). The requirements of our analyses skew the data a bit toward persons who stay together for several years.

The models specify the probability of each of the outcome variables: partisan support, preference, and constancy, for each of the two dominant parties in each country, the Social Democrats and Christian Democrats/Socials in Germany and Labour and the Conservatives in Britain and turnout and vote choice in Britain.

EMPIRICAL GENERALIZATIONS ABOUT PARTISANSHIP WITH ROBUST SUPPORT

1. The citizenry chooses between one or the other of the major parties.
2. During the years of the surveys, there is a decline in the aggregate levels of partisan support in each country. Young people are especially slow to claim to support political parties.
3. On average for each person, the rate of preference constancy for Party A/B is 0.40–0.60; the probability of ever choosing Party B/A is less than 0.10, and the probability of ever naming no party is 0.25–0.30.
4. As a result, the probability of naming A/B is not the complement of preferring B/A. Most people decide between naming A/B and selecting no political party. People are bounded partisans.
5. There is a statistically significant correlation between the predictors of partisan support and choice.

GENERAL THEORETICAL PROPOSITIONS

Ancient wisdom, research schools across the social sciences, and neurological studies agree that humans, *qua* humans, send and receive messages. They influence each other. The more they interact, the more messages are exchanged, the greater the level of trust, and the greater the level of recurrent learning, the greater is the level of reciprocal influence.

HYPOTHESES, FOR WHICH OUR ANALYSES PROVIDE ROBUST EMPIRICAL SUPPORT

1. The mix of partisan preferences among members of a person's household directly affects the probability of choosing Party A/B, net of all other predictor variables. To be precise, moving along a

Partisan Families

scale that extends from a net positive of two to a net negative of two changes the probability of choice from a very high probability of naming the party to a virtual certainty of not selecting the party.

2. Political interest has a strong positive effect on the probability of partisan support and turnout, net of all other predictor variables. It has no effect on partisan choice or the reciprocal partisan relationships within households.
3. The broad social contexts of social class and religion remain important to partisan choice and constancy and to electoral choice, even as there is evidence of some decline in the centrality in most people's lives.
4. Exposure to elections by itself has no effect on the probability of partisan support.
5. Wives/mothers stand at the heart of the exchange of political preferences within households. They positively affect and are affected by husbands/fathers and children. Fathers and children do not have a consistent partisan relationship with each other.
6. Partisan agreement in couples is autoregressive: the more they share the same social locations and identification and the more years they live together, the higher is the rate of partisan agreement.
7. Partisan choice and constancy combine with the electoral preferences of others in the household to account for the probability of turnout and whether or not a person casts a ballot for Party A/B.
8. Variations in political interest among members of the household do not affect relative political influence within the household.

THEORETICAL IMPLICATIONS

1. People create a choice set from the list of available political parties.
2. Contrary to the political scientists who first apply the behavioral revolution to the study of partisanship (see Chapter 1), there is no need to choose between the analysis of the family and the analysis of political parties.
3. Elaborating on Simon (see Chapter 1), mechanisms that sustain as well as initiate behavior are external to the individual.
4. Asymmetries in political interest do not underpin political influence in households.
5. What accounts for the household effect on partisanship? Family members have exceptionally high levels of interaction and exceptionally high levels of trust and affection, and so they have exceptionally powerful effects on each member's probability of partisan support, choice, and constancy.

Conclusion

6. Why do children's politics resemble more closely the politics of their mothers than of their fathers? Again, the frequency of interaction and the level of trust and affection, as well as prior learning, provide the answer. They spend more time with their mothers; they learn more from them, and they teach them as well.
7. The frequency of interaction affects past learning and the probability of current and future social influence.
8. The frequency of interaction affects the probability of trust and emotional bonds between and among persons. These ties increase the probability of social influence.
9. We expect that differential political mobilization by the political parties reinforces these relationships, though our survey data do not allow us to test this claim. Because campaigns seek to mobilize persons who are likely to support Party A/B, they reach persons whose intimate social ties induce them to support that party. The strategic choices of party leaders reinforce the bounded party system that each person constructs and the likelihood that each person remains a bounded partisan.
10. The result is a reciprocal relationship between the social logic of partisanship and the behavioral effects of the strategic calculations of party leaders (affirming an insight offered and then set aside by Campbell, Converse, Miller, Stokes, Eulau, and Key – see Chapter 1).
11. Reciprocal relationships, as opposed to one-way causal flows, characterize many political relationships.

As humans, we express and interpret views, feelings, knowledge, wishes, and the like. As we do this, we influence each other; we learn from each other. As Onkeles and Aristotle taught centuries ago and as so many have understood since, communication is inherent to the meaning of human.

Appendix

THE SURVEYS

Full descriptions of the German Socio-Economic Panel Study and the British Household Panel Survey, the samples, questions, constructed variables, and weighting variables may be obtained from the respective Web sites; for Deutsche Institut für Wirtschaftsforschung (DIW Berlin) go to http://www.diw.de/GSOEP, for the Institute for Social and Economic Research at the University of Essex, go to http://www.iser.essex.ac.uk.

We describe the measures of partisan support, choice, and constancy in the Preface. Some measures (age, partisan preference in the region of residence, the Goldthorpe categories of occupation, household income, education category) are readily interpreted. Other technical matters need elaboration. Some variables like the measures of partisanship are asked each year. When we use a variable that is asked intermittently, for example religious identification, we impute forward the first response until the next time that it is asked. Omitting it from the analysis would skew and drastically reduce the size of the two samples.

REGARDING THE DATA TAKEN FROM GSOEP

A German is a person who (a) was born in Germany, (b) has only held German citizenship, and (c) lived in the Federal Republic of Germany (FRG) in 1989.

We use the English translation of GSOEP's survey in order to describe the measures. The following question defines political interest: "First of all in general: How interested are you in politics?" Then the respondents are offered the following choices: "Very interested [4 points], fairly interested [3 points], not very interested [2 points], and not interested [1 point]."

Religion is measured by a question that asks about "membership in a church or denomination." It allows for five options: Roman Catholic,

Appendix

Protestant, Other Christian, non-Christian, and no religion. Frequency of church attendance derives from the question: "Which of the following activities do you do in your free time? Go to church, attend other religious events.... Please enter how often you practice each activity: each week [3 points]; each month [2 points]; less often [1 point]; never [0 points]."

Union membership derives from two related questions: "Are you a member of a workplace union/of any union?"

Questions on volunteer work and social contact are found in the context of how persons spend their time and are asked almost every second year: "Now some questions about your free time. How frequently do you do the following activities? Volunteer work in clubs, associations, or social services; visit with friends, relatives, or neighbors." Answers range from "never to weekly." Because the time-use variables are overdispersed on the category "never," we recoded them to binary variables of "no = never" and "yes = else."

Economic problems are tapped by two questions: "What about the following areas: Do they worry you? If employed, the security of your job? General economic condition?" The scale varies among very worried [2 points], slightly worried [1 point], not worried [0 points]. Because of problems of multicollinearity, we excluded responses to questions about personal financial circumstances from our models.

REGARDING DATA TAKEN FROM BHPS

The following question taps political interest: "How interested would you say you are in politics? Would you say you are: Very interested, fairly interested, not very interested, not at all interested?" The measure awards points for the level of political interest: three for very interested, two for fairly interested, one for not very interested, and none for not at all interested. We added one point to each response so that the BHPS and GSOEP scales would be the same.

Religion: "Do you regard yourself as belonging to any particular religion? If Yes, which?" We collapsed the various options into seven categories. They are Church of England; Other Protestant; Roman Catholic; Jewish; Muslim, Other, and none. Religious attendance: How often, if at all, do you attend religious services or meetings? "Practically Never;" "At least once a year;" "At least once a month;" "Once a week or more?"

The social class questions first ask about identity: "Do you think of yourself as belonging to any particular social class?" A second question offers two options: seven categories of working class and middle class, which we have collapsed into two – working class and middle class.

A positive answer to "Are you currently a member of a Trades Union?" labels a person as a union member.

Appendix

The following question measures activity in social organizations: "Do you join in the activities of any of the organizations on this card on a regular basis?"

The measure of economic problems taps income issues by asking whether the respondents have had problems paying for housing and the number of times that they answered that their economic conditions "were now worse."

THE MODELS

Models are more or less useful for the problem at hand. When questions are complex, as in our analysis where we expect to find two- and three-stage, reciprocal, and endogenous relationships, there is almost never a perfect model; each has strengths and weaknesses. The Heckman Probit Selection model enables us to approach partisan choice and voting in a way that recognizes their inherent two-stage process and to control for potential problems associated with selection bias that derive from differential responses to the question on partisan support. It does not allow us to model directly reciprocal influences within households. The Zero-Inflated Negative Binomial model also includes two steps: which party and how frequently it is named. However, it may be applied only to the subset of respondents who are in all waves and it does not allow us to assess directly household effects. We use instrumental variable probit analysis to examine the reciprocal relations by couples and three-stage least squares linear probability models to examine interpersonal relationships among household triads. These models, however, can analyze only one dependent variable at a time, and we analyze partisan preference and vote choice.

Heckman Probit Selection Model

This approach models the two-step decision process that links partisan support and choice. It also accounts for problems of selection bias that follow when persons who deny that they support a party are removed from the analysis. To elaborate, if cases that are *non*randomly selected on the dependent variable (in this case party choice) are excluded from the analysis, selection bias occurs, and subsequent models are misspecified. This results in biased parameter estimates and misleading substantive inferences. The Heckman Probit Selection model also offers a summary statistic that describes the strength of association between the answer to the first and second questions. The rho statistic is the correlation between the error terms of equation (a) and (b). A significant coefficient indicates a relationship between the process to pick a party and the process to name a specific one. Although a selection effect does exist in each of the

Appendix

models estimated, the parameter estimates for the Heckman Probit Selection models do not differ markedly from simple standard probit models. These standard probit models are not shown but can be obtained from the authors. As a result, each table contains two sets of models: an analysis of the decision to support a party and the choice of party (and see Greene 1997; Heckman 1979; and Wooldridge 2003 for more detailed descriptions of Heckman selection models).

The Zero Inflated Negative Binomial (ZINB) Model

Because the count results presented in Tables 2.7 and 2.8 indicate that there is an excess of zeros and because the data are overdispersed (i.e., the variance of the dependent variable in each case is greater than the mean), we employ the Zero-Inflated Negative Binomial model. Were we to use a simple count model (such as the Poisson model), we would seriously underpredict the number of zeros. The ZINB model accounts for this underprediction by adjusting both the conditional variance and the conditional mean of the dependent variable. It adjusts "the mean structure to explicitly model the production of zero counts" (Long 1997, 242). To elaborate, the ZINB model assumes that zeros are generated by a process that is different from the one that generates positive counts. Indeed, two different processes (one binary and the other a count process) generate the zeros. Each respondent is hypothesized to belong to one of two groups, membership in which is determined by the appropriateness of the two processes. Using an extra parameter estimate, the ZINB model examines explicitly whether an individual belongs to either the binary or count group. This outcome is a function of the respondent's characteristics. Note that the Negative Binomial Regression model adjusts for *only* the conditional variance. Postestimation specification tests, with accompanying statistics (which we show in the tables) alert us to the goodness of fit of the respective count models. In all the models presented, the ZINB outperforms all other count models. Not only is this model appropriate on technical grounds, but the model's underlying logic and the data-generating processes it is meant to depict are appropriate to a theory of bounded partisanship. Some individuals situate themselves on one side of the partisan hedge and never support the other party. These people should be modeled according to a binary process (and see Long 1997 and Winkelmann 2000 on ZINB models).

Reciprocal Models

In reciprocal models, we use instrumental variables to assess partisan choice or electoral decisions for others in the households. They are derived

Appendix

from the results of the initial Heckman Probit Selection models and the results of logit equations for each type of respondent. The instrumental variables display statistically significant relationships with the dependent variable but do not correlate with other predictor variables. As we note in Chapter 5, Heckman and Macurdy (1985) and Heckman and Snyder (1997) justify the application of least squares regression analysis to a binary dependent variable. Instrumental variable probit analysis does not permit the analysis of three-way reciprocal effects. The outcomes of a series of two-stage instrumental probit analyses substantially mimic those obtained from the three-stage OLS regression model. We ran a series of boot-strap simulations, using the parameter estimates from these models in order to determine the predicted probabilities of each dependent variable. These use the complete range of actual values on the predictor variables. The 95 percent confidence intervals of the point estimates of the predicted probabilities always fall between zero and one, providing overwhelming empirical justification for the application of this model to our data.

Additional technical matters flow from the use of household panel survey data. Because the surveys sample their respondents by household, the respondents cluster together. As a result, the observations are not independent of each other. This reduces the estimated size of the standard error, inflates the Z-values in the models, and may imply the presence of statistical significance when it is not there. The models provide several ways to avoid this problem: sometimes they offer a "cluster" command, which corrects the standard errors and, therefore, the Z-scores, and in some of the logistic regressions, we are able to apply the xtlogit command, which controls for the autocorrelation that affects cross-sectional analyses of time-series data. Also, some of our analyses include all persons who are ever interviewed; some of these many times and some only once. Again, xtlogit enables us to address this matter directly.

POSTESTIMATION PROBABILITIES AND COUNTS

Postestimation through boot-strapping of parameter estimates enables us to calculate the expected values of the dependent variable given various combinations of values of the main predictor variables and the parameter estimates from the regression equations.

For some models, Stata/SE 8.2 offers the prvalue command. Where it does not (e.g., in the Probit-Heckman Selection models, the two-stage instrumental probit models, and the three-stage linear probability models), we program the simulations ourselves. While we could simply calculate the point estimate of the expected values of each dependent variable, this would ignore the existence of the error term in the regression

Appendix

equations. As a result, we have little idea of how confident we are that the point estimates we have calculated are correct. To overcome this problem, we perform boot-strap calculations of the expected values of the dependent variables using information about the parameter estimates and the covariance matrix. We then calculate 1,000 sets of new parameter estimates, set our main independent variables at their desired values while setting all others at the means, and thereby calculate 1,000 estimates of the expected values of each of the dependent variables. We are then able to determine the 95 percent confidence intervals for the expected values of the dependent variables with ease. These confidence intervals are plotted on the figures as vertical bands.

References

Achen, Christopher. 1975. "Mass Political Attitudes and the Survey Response." *American Political Science Review* 69: 1218–31.
  1992. "Social Psychology, Demographic Variables, and Linear Regression: Breaking the Iron Triangle." *Political Behavior* 14: 195–211.
  2002. "Parental Socialization and Rational Party Identification." *Political Behavior* 24: 151–70.
Acock, Alan C., and Vern L. Bengtson. 1978. "On the Relative Influence of Mothers and Fathers: A Covariance Analysis of Political and Religious Socialization." *Journal of Marriage and the Family* 40(3): 519–30.
Alford, C. Fred. 1994. *Group Psychology and Political Theory*. New Haven, Conn.: Yale University Press.
Anderson, Christopher J., Silvia M. Mendes, and Yuliya V. Tverdova. 2004. "Endogenous Economic Voting: Evidence from the 1997 British Election." *Electoral Studies* 20: 1–26.
Anderson, Christopher J., and Aida Paskeviciute. 2004. "Macro-Politics and Micro-Behavior: Mainstream Politics and the Frequency of Political Discussion in Contemporary Democracies." In *The Social Logic of Politics*, Alan S. Zuckerman, ed. Philadelphia: Temple University Press: 228–50.
Aquinas, Thomas. 1964. *Commentary on the Nicomechean Ethics*, tr. C. I. Litzinger, O. P. Chicago: Henry Regnery, Library of Living Catholic Thought. II: 855.
  1983. *Treatise on Happiness*. Question 4, Article 8. South Bend Ind.: University of Notre Dame Press: 52.
Aristotle. 1962. *Nicomechean Ethics*. New York: Bobbs-Merrill, tr. Martin Oswald.
Aron, Art, and Elaine Aron. 2000. "Self-expansion Motivation and Including Other in the Self." In *The Social Psychology of Personal Relationship*, William Ickes and Steve Duck, eds. New York: Wiley: 109–28.
Arrow, Kenneth. 1986. "Rationality of Self and Others in an Economic System." *The Journal of Business* 59: S385–S398.
Aumann, Robert J. 2005. "Consciousness." Discussion Paper 391, Center for the Study of Rationality, Hebrew University of Jerusalem, May.
Axelrod, Robert. 1997a. "The Dissemination of Culture: A Model with Local Convergence and Global Polarization." *Journal of Conflict Resolution* 41: 203–26.

References

1997b. *The Complexity of Cooperation: Agent-Based Models of Competition and Collaboration.* Princeton, N.J.: Princeton University Press.

Baker, Kendall. 1974. "The Acquisition of Partisanship in Germany." *American Journal of Political Science* 18(3): 569–82.

Bannon, Declan. 2003. "Market Segmentation and Political Marketing." Political Science Association Conference, PMG Panel. Lincoln, England.

Bantle, Christian, and John P. Haisken-DeNew. 2002. "Smoke Signals: The Intergenerational Transmission of Smoking Behavior." German Institute for Economic Research-DIW.

Barker, Ernest, ed. and tr. 1962. *The Politics of Aristotle.* New York: Oxford University Press.

Bartle, John. 1999. "Improving the Measurement of Party Identification." In *British Elections and Parties Review Vol. 9*, Justin Fisher, Philip Cowley, David Denver and Andrew Russell, eds. London: Frank Cass: 119–35.

2001. "The Measurement of Party Identification in Britain: Where Do We Stand Now?" In *British Elections & Parties Review Vol. 11*, Jon Tonge, Lyn Bennie, David Denver and Lisa Harrison, eds. London: Frank Cass: 1–14.

2003. "Focus Groups and Measures of Party Identification: An Exploratory Study." *Electoral Studies* 22: 217–37.

Beck, Paul Allen, Russell J. Dalton, Steven Greene, and Robert Huckfeldt. 2002. "The Social Calculus of Voting: Interpersonal, Media, and Organizational Influences on Presidential Choices." *American Political Science Review* 96(1): 57–73.

Beck, Paul Allen, and M. Kent Jennings. 1975. "Parents as 'Middlepersons' in Political Socialization." *Journal of Politics* 37(1): 83–107.

1991. "Family Traditions, Political Periods, and the Development of Political Orientations." *Journal of Politics* 53: 742–63.

Becker, Gary. 1964. *Human Capital.* New York: Columbia University Press.

1981. *A Treatise on the Family.* Cambridge, Mass.: Harvard University Press.

Becker, Gary. 1985. "Human Capital, Effort, and the Sexual Division of Labor." *Journal of Labor Economics* 3: S33–S58.

Berelson, Bernard R., Paul F. Lazarsfeld, and William N. McPhee. 1954. *Voting: A Study of Opinion Formation in a Presidential Campaign.* Chicago: University of Chicago Press.

Bergmann, Knut, and Wolfram Wickert. 1999. "Selected Aspects of Communication in German Election Campaigns." In *Handbook of Political Marketing*, Bruce I. Newman, ed. Thousand Oaks, Calif.: Sage: 455–84.

Berns, Gregory S., Jonathan Chappelow, Caroline F. Zink, Giuseppe Pagnoni, Megan E. Martin-Skurski, and Jim Richards. 2005. "Neurobiological Correlates of Social Conformity and Independence During Mental Rotation." *Biological Psychiatry* 58: 245–53.

Beyer, Mary Alice, and Robert Whitehurst. 1976. "Value Change with Length of Marriage: Some Correlate of Consonance and Dissonance." *International Journal of Marriage and the Family* 6: 109–20.

Bion, W. R. *Experiences in Groups and Other Papers.* 1961. New York: Basic Books.

Blais, André. 2000. *To Vote or Not to Vote: The Merits and Limits of Rational Choice Theory.* Pittsburgh: University of Pittsburgh Press.

Blais, André, Elisabeth Gidengil, Richard Nadeau, and Neil Nevitte. 2001. "Measuring Party Identification: Britain, Canada, and the United States." *Political Behavior* 23: 5–21.

References

Bottomore, T. B., and Maximilien Rubel, eds. 1956. *Karl Marx: Selected Writings in Sociology and Social Philosophy.* New York: McGraw-Hill.

Boudon, Raymond. 1992. "Subjective Rationality and the Explanation of Social Behavior." In *Economics, Bounded Rationality and the Cognitive Revolution*, Herbert Simon, ed. Brookfield, Vt.: Edward Elgar.

——— 1998. "Social Mechanisms without Black Boxes." In *Social Mechanisms: An Analytical Approach to Social Theory*, Peter Hedström and Richard Swedberg, eds. New York: Cambridge University Press.

Brynin, Malcolm, and David Sanders. 1997. "Party Identification, Political References, and Material Conditions: Evidence from the British Household Panel Survey, 1991–2." *Party Politics* 3: 53–77.

Butler, David E., and Donald Stokes. 1969. *Political Change in Britain: The Evolution of Party Choice.* London: Macmillan.

——— 1974. *Political Change in Britain: The Evolution of Electoral Choice.* New York: St. Martin's Press.

Campbell, Angus, Philip E. Converse, Warren. E. Miller, and Donald. E. Stokes. 1960. *The American Voter.* New York: Wiley.

——— 1966. *Elections and the Political Order.* New York: Wiley.

Cartwright, Dorwin, ed. 1964 [1951]. *Field Theory in Social Science: Selected Theoretical Papers by Kurt Lewin.* New York: Harper and Row.

Coile, Courtney C. 2003. *Retirement Incentives and Couples' Retirement Decisions.* Working Paper 9496; http://www.nber.org/papers/w9496. National Bureau of Economic Research, Cambridge, Mass. (December).

Collins, W. Andrew, and Brett Laursen. 2000. "Adolescent Relationships: The Art of Fugue." In *Close Relationships: A Sourcebook*, Clyde Hendrick and Susan Hendrick, eds. Thousand Oaks, Calif.: Sage Publications: 59–70.

——— 2004. "Parent-Adolescent Relationships and Influences." In *Handbook of Adolescent Psychology*, Richard M. Lerner and Laurence Steinberg, eds. Hoboken, N.J.: Wiley: 331–61.

Coltrane, Scott. 2000. "Research on Household Labor Modeling and Measuring the Social Embeddedness of Routine Family Work." *Journal of Marriage and the Family* 62(4): 1208–33.

Conover, Pamela J., Donald D. Searing, and Ivor M. Crewe. 2002. "The Deliberative Potential of Political Discussion." *British Journal of Political Science* 32(1): 21–62.

Converse, Philip E. 1964. "The Nature of Belief Systems in Mass Publics." In *Ideology and Discontent.* David Apter, ed. New York: The Free Press: 206–61.

——— 1969. "Of Time and Partisan Stability." *Comparative Political Studies* 2(4): 139–71.

——— 1976. *The Dynamics of Party Support: Cohort-Analyzing Party Identification.* Beverly Hills, Calif.: Sage.

Cooke, Lynn. 2004. "The Gendered Division of Labor and Family Outcomes in Germany." *Journal of Marriage and the Family* 66: 1246–59.

Crewe, Ivor. 1980. "Negative Partisanship: Some Preliminary Ideas Using British Data." Paper prepared for the Planning Session on Problems in Comparative Survey Research in Political Behavior: Issues in Data Collection and Analysis, Joint Sessions of the European Consortium for Political Research, Florence, March.

Cunningham, Michael R., and Anita P. Barbee. 2000. "Social Support." In *Close Relationships: A Sourcebook*, Clyde Hendrick and Susan S. Hendrick, eds. Thousand Oaks, Calif.: Sage: 273–86.

References

Dalton, Russell J. 2000. "The Decline of Party Identifications." In *Parties Without Partisans: Political Change in Advanced Industrial Democracies*, Russell J. Dalton and Martin P. Wattenberg, eds. New York: Oxford University Press: 19–36.

Dalton, Russell J., and Wilhelm Bürklin. 1996. "The Two German Electorates." In *Germans Divided*, Russell Dalton, ed. Oxford: Berg: 183–208.

2003. "Wähler als Wandervögel: Dealignment and the German Voter." *German Politics and Society* 21: 57–75.

Dalton, Russell J., and Martin P. Wattenberg, eds. 2000. *Parties Without Partisans: Political Change in Advanced Industrial Democracies*. New York: Oxford University Press.

Davies, James C. 1970. "The Family's Role in Political Socialization." In *Learning about Politics*, Roberta S. Sigel, ed. New York: Random House.

De Graaf, Nan Dirk, and Anthony Heath. 1992. "Husbands' and Wives' Voting Behaviour in Britain: Class-dependent Mutual Influence of Spouses." *Acta Sociologica* 35(4): 311–22.

Denver, David, and Gordon Hands. 2002. "Post-Fordism in the Constituencies: The Continuing Development of Constituency Campaigning in Britain." In *Do Political Campaigns Matter? Campaign Effects in Elections and Referendums*, David M. Farrell and Rüdiger Schmitt-Beck, eds. London and New York: Routledge: 108–26.

Denver, David, Gordon Hands, Justin Fisher, and Ian MacAllister. 2003. "Constituency Campaigning in Britain 1992–2001." *Party Politics* 9(5): 541–59.

Denver, David, Gordon Hands, and Simon Henig. 1998. "Triumph of Targeting? Constituency Campaigning in the 1997 Election." In *British Elections and Parties Review* VIII, David Denver, Justin Fisher, Philip Cowley, and Charles Pattie, eds. London: Cass: 171–90.

Denver, David, Gordon Hands, and Ian MacAllister. 2004. "The Electoral Impact of Constituency Campaigning in Britain." *Political Studies* 52: 289–306.

Diamond, Gregory Andrade. 2001. "Implications of a Latitude-Theory Model of Citizen Attitudes for Political Campaigning, Debate, and Representation." In *Citizen Politics: Perspectives from Political Psychology*, James H. Kuklinski, ed. New York: Cambridge University Press: 289–312.

Dogan, Mattei. 1967. "Political Cleavage and Social Stratification in France and Italy." In *Party Systems and Voter Alignments: Cross-National Perspectives*, Seymour M. Lipset and Stein Rokkan, eds. New York: The Free Press: 129–96.

Douglas, Mary. 1986. *How Institutions Think*. Syracuse, N.Y.: Syracuse University Press.

Douglas, Mary, and Steven Ney. 1998. *Missing Persons: A Critique of Personhood in the Social Sciences*. Berkeley, Calif.: University of California Press.

Downs, Anthony. 1957. *An Economic Theory of Democracy*. New York: Harper and Row.

Durkheim, Emile. 1966 [1933]. *The Division of Labor in Society*. New York: The Free Press, tr. George Simpson.

Duverger, Maurice. 1967. *Political Parties*. New York: Wiley, tr. Robert C. North.

Easton, David, and Jack Dennis. 1976. *Children and the Political System: Origins of Political Legitimacy*. New York: McGraw Hill.

Edlund, Lena, Rohini Pander, and Lailer Haider. 2005. "Unmarried Parenthood and Redistributive Politics." *The Journal of the European Economic Association* 3: 95–119.

References

Elster, Jan. 1998. "A Plea for Mechanisms." In *Social Mechanisms: An Analytical Approach to Social Theory*, Peter Hedström and Richard Swedberg, eds. New York: Cambridge University Press: 45–73.

Erikson, Robert S. 2004. "Macro vs. Micro-Level Perspectives on Economic Voting: Is the Micro-Level Evidence Endogenously Induced?" Political Methodology Meetings, Stanford University, July.

Erikson, Robert S., Michael B. MacKuen, and James A. Stimson. 2002. *The Macro Polity*. New York: Cambridge University Press.

Eulau, Heinz. 1962. *Class and Party in the Eisenhower Years: Class Roles and Perspectives in the 1952 and 1956 Elections*. New York: The Free Press of Glencoe.

  1980. "The Columbia Studies of Personal Influence." *Social Science History* 4: 207–28.

  1986. *Politics, Self, and Society*. Cambridge, Mass.: Harvard University Press.

Euromonitor International Web site: *http://www.euromonitor.com/Marketing_to_children*.

Eurostat. 2003. *Time Use at Different Stages of Life: Results from 13 Countries*. Luxembourg: Office of Publications of the European Communities.

  2004. *How Europeans Spend Their Time: Everyday Life of Women and Men*. Luxembourg: Office for Official Publications of the European Communities.

Evans, Geoffrey, ed. 1999a. *The End of Class Politics? Class Voting in Comparative Context*. New York: Oxford University Press.

  1999b. "Class Voting: From Obituary to Appraisal." In *The End of Class Politics? Class Voting in Comparative Context*, Geoffrey Evans, ed. New York: Oxford University Press: 1–20.

  1999c. "Class and Vote: Disrupting the Orthodoxy." In *The End of Class Politics? Class Voting in Comparative Context*, Geoffrey Evans, ed. New York: Oxford University Press: 323–4.

Evans, Geoffrey, and Robert Andersen. 2001. "Endogenizing the Economy: Political Preferences and Economic Perceptions Across the Electoral Cycle." CREST Working Paper 88. Available on *http://www.crest.ox.ac.uk*.

Falter, J., H. Schoen, and C. Caballero. 2000. "Zur Validierung des Konzepts 'Parteiidentifikation' in der Bundesrepublik." In *50 Jahre empirische Wahlforschung in Deutschland. Entwicklung, Befunde, Perspektiven, Daten*, M. Klein, W. Jagodzinski, E. Mochmann, and D. Ohr. Wiesbaden, eds. Westdeutscher Verlag: 235–71. ["On the Validity of the Concept of Party Identification in the Federal Republic."]

Farrell, David M., and Rüdiger Schmitt-Beck. 2002. *Do Political Campaigns Matter? Campaign Effects in Elections and Referendums*. London and New York: Routledge.

Finkel, Steven E. 1993. "Reexamining the 'Minimal Effects' Model in Recent Presidential Campaigns." *Journal of Politics* 55(1): 1–21.

Finkel, Steven E., and Peter R. Schrott. 1995. "Campaign Effects on Voter Choice in the German Election of 1990." *British Journal of Political Science* 25: 349–77.

Fiorina, Morris P. 1981. *Retrospective Voting in American National Elections*. New Haven, Conn.: Yale University Press.

  2002. "Parties and Partisanship: A 40-Year Retrospective." *Political Behavior* 24: 93–115.

Fisher, Kimberly, Andrew McCulloch, and Jonathan Gershuny. 1999. *British Fathers and Children: A Report for Channel 4 "Dispatches."* Institute for

References

Social and Economic Research, University of Essex, Wivenhoe Park, Essex, Colchester CO4 3SQ. Available on *http://www.iser.essex.ac.uk*.

Fowler, James H. 2005. "Turnout in a Small World." In *The Social Logic of Politics*, Alan S. Zuckerman, ed. Philadelphia: Temple University Press: 269–88.

Franklin, Mark N., Tom Mackie, and Henry Valen. 1992. *Electoral Change: Responses to Evolving Social and Attitudinal Structures in Western Countries*. New York: Cambridge University Press.

Gauthier, Anne H., Timothy Smeeding, and Frank F. Furstenberg, Jr. 2004. "Do We Invest Less Time in Children? Trends in Parental Time in Selected Industrialized Countries since the 1960s." *Population and Development Review*. 30(4): 579–624.

Gigerenzer, Gerd, and Reinhard Selten, eds. 2001a. *Bounded Rationality: The Adaptive Toolbox*. Cambridge, Mass.: MIT Press.

2001b. "Rethinking Rationality." In *Bounded Rationality: the Adaptive Toolbox*, Gerd Gigerenzer and Reinhard Selten, eds. Cambridge, Mass.: MIT Press: 1–12.

Gimpel, James, and J. Celeste Lay. 2005. "Partisan Identification, Local Partisan Contexts, and the Acquisition of Participatory Attitudes." In *The Social Logic of Politics*, Alan S. Zuckerman, ed. Philadelphia: Temple University Press: 209–28.

Glaser, William A. 1959–60. "The Family and Voting Turnout." *Public Opinion Quarterly* 23(4): 563–70.

Gluchowski, P. M., and U. von Wilamowitz-Moellendorff. 1998. "The Erosion of Social Cleavages in Western Germany, 1971–97." In *Stability and Change in German Elections: How Electorates Merge, Converge, or Collide*, Christopher J. Anderson and Carsten Zelle, eds. Westport, Conn.: Praeger: 13–32.

Goldthorpe, John H. 1999a. "Modeling the Pattern of Class Voting in British Elections: 1964–1992." In *The End of Class Politics? Class Voting in Comparative Context*," Geoffrey Evans, ed. New York: Oxford University Press: 59–82.

1999b. "Critical Commentary: Four Perspectives on the End of Class Politics." In *The End of Class Politics? Class Voting in Comparative Context*," Geoffrey Evans, ed. New York: Oxford University Press: 318–22.

Gonzalez, Richard, and Dale Griffin. 2000. "On the Statistics of Interdependence: Treating Dyadic Data with Respect." In *The Social Psychology of Personal Relationships*, William Ickes and Steve Duck, eds. New York: Wiley: 181–213.

Granovetter, Mark. 1973. "The Strength of Weak Ties." *American Journal of Sociology* 78: 1360–80.

1974. *Getting a Job: A Study of Contacts and Careers*. Cambridge, Mass.: Harvard University Press.

Gray, Mark. 2003. "In the Midst of Fellows: The Social Context of the American Turnout Decision." Paper prepared for and presented at the Meetings of the American Political Science Association, Philadelphia, August 28–3.

Green, Donald, Bradley Palmquist, and Eric Schickler. 2002. *Partisan Hearts and Minds: Political Parties and the Social Identities of Voters*. New Haven, Conn.: Yale University Press.

Greene, Steven. 2002. "The Social-Psychological Measurement of Partisanship." *Political Behavior* 24: 171–97.

References

Greene, W. H. 1997. *Econometric Analysis*, 3rd ed. upper Saddle River, N.J. Prentice Hall.
Greenstein, Fred I. 1965. *Children and Politics*. New Haven, Conn.: Yale University Press.
Haslam, S. Alexander, Craig McCarty, and John C. Turner. 1996. "Salient Group Memberships and Persuasion: The Role of Social Identity in the Validation of Beliefs." In *What's Social About Social Cognition?: Research on Socially Shared Cognition in Small Groups*, Judith L. Nye and Aaron Brower, eds. Thousand Oaks, Calif.: Sage: 29–58.
Hays, Bernadette C., and Clive S. Bean. 1992. "The Impact of Spousal Characteristics on Political Attitudes in Australia." *Public Opinion Quarterly* 56: 524–9.
Heckman, James. 1979. "Sample Selection Bias as a Specification Error." *Econometrica* 45: 153–61.
Heckman James J., and Thomas E. Macurdy. 1985. "A Simultaneous Equations Linear Probability Model." *Canadian Journal of Economics* 18(1): 28–37.
Heckman, James J., and James N. Snyder, Jr. 1997. "Linear Probability Models of the Demand for Attributes with an Empirical Application to Estimating the Preferences of Legislators." *Rand Journal of Economics* 28: 142–89.
Heineman, Elizabeth D. 1999. *What Difference Does a Husband Make: Women and Marital Status in Nazi and Postwar Germany*. Berkeley: University of California Press.
Hendrick, Clyde, and Susan S. Hendrick, eds. 2000. *Close Relationships: A Handbook*. Thousand Oaks, Calif.: Sage.
Hernes, Gunnar. 1998. "Virtual Reality." In *Social Mechanisms: An Analytical Approach to Social Theory*, Peter Hedström and Richard Swedberg, eds. New York: Cambridge University Press: 74–101.
Hess, Robert D., and Judith V. Torney. 1968. *The Development of Political Attitudes in Children*. Anchor Books; Garden City, N.Y.: Doubleday.
Holbrook, Thomas M., and Scott D. McClurg. 2005. "The Mobilization of Core Supporters: Campaigns, Turnout and Electoral Composition in the United States." *American Journal of Political Science* 49(4): 689–703.
Holt, Robert T., and John Turner. 1968. *Political Parties in Action*. New York: Free Press.
Holzhacker, Ronald D. 1999. "Campaign Communication and Strategic Responses to Change in the Electoral Environment." *Party Politics* 5(4): 439–69.
Huckfeldt, Robert, Paul Allen Beck, Russell J. Dalton, and Jeffrey Levine. 1995. "Political Environments, Cohesive Social Groups, and the Communication of Public Opinion." *American Journal of Political Science* 39(4): 1025–54.
Huckfeldt, Robert, Paul E. Johnson, and John Sprague. 2004. *Political Disagreement: The Survival of Diverse Opinions within Communication Networks*. New York: Cambridge University Press.
2005. "Individuals, Dyads, and Networks: Autoregressive Patterns of Political Influence." In *The Social Logic of Politics*, Alan S. Zuckerman, ed. Philadelphia: Temple University Press: 21–50.
Huckfeldt, Robert, and John Sprague. 1995. *Citizens, Politics, and Social Communication: Information and Influence in an Election Campaign*. New York: Cambridge University Press.
Huddy, Leonie. 2003. "Group Identity and Political Cohesion." In *Oxford Handbook of Political Psychology*, David O. Sears, Leonie Huddy and Robert Jervis, eds. New York: Oxford University Press: 511–58.

References

Huston, Ted L., John P. Caughlin, and Robert Houts. 2000. "How Does Personality Matter in Marriage? An Examination of Trait Anxiety, Negativity, and Marital Satisfaction." *Journal of Personality and Social Psychology* 78: 326–36.

Hyman, Herbert. 1959. *Political Socialization*. New York: Free Press.

Ickes, William, and Steve Duck, eds. 2000. *The Social Psychology of Personal Relationships*. New York: Wiley.

Iversen, Torben, and Frances Rosenbluth. 2003. "The Political Economy of Gender: Explaining Cross-National Variation in the Gender Division of Labor and the Gender Voting Gap." Paper presented to the Annual Meetings of the American Political Science Association, Philadelphia, August 28–31.

Jaros, Dean. 1973. *Socialization to Politics*. New York: Praeger.

Jennings, M. Kent, and Richard G. Niemi. 1968. "The Transmission of Political Values from Parent to Child." *American Political Science Review* 62: 169–84.

1981. *Generations and Politics: A Panel Study of Young Adults and their Parents*. Princeton, N.J.: Princeton University Press.

Joesch, Jutta M., and C. Katharina Spiess. 2002. "European Mothers' Time with Children: Differences and Similarities across Nine Countries." Paper prepared for the IZA and CIM Workshop, *The Future of Family and Work: Evaluation of Family Friendly Policies*. IZA, Bonn and CIM, Aarhus. Bonn. May 10–12. Available on http://www.oza.prg/en/papers/iza_cim_joesch_spiess.pdf.

Johnson, Paul E., and Robert Huckfeldt. 2005. "Agent-Based Explanations for the Survival of Disagreement in Social Networks." In *The Social Logic of Politics*, Alan S. Zuckerman, ed. Philadelphia: Temple University Press.

Johnson, Richard W., and Melissa M. Favreault. 2001. *Retiring Together or Working Alone: The Impact of Spousal Employment and Disability on Retirement Decisions*. Center for Retirement Education at Boston College. Available on http://www.bc.edu/crr.

Johnston, Ron J. 1999. "Context, Conversation and Conviction: Social Networks and Voting at the 1997 British General Election." *Political Studies* 47: 877–89.

Johnston, Ron J., and Charles Pattie. 2005. "Putting Voters in their Places: Local Context and Voting in England and Wales." In *The Social Logic of Politics*, Alan S. Zuckerman, ed. Philadelphia: Temple University Press: 184–208.

Johnston, Ron J., Rebecca Sarker, Kelvyn Jones, Anne Bolster, Carol Propper, and Simon Burgess. 2005. "Egocentric Economic Voting and Changes in Party Choice: Great Britain 1992–2001." *Journal of Elections, Public Opinion, and Parties* 15: 129–44.

Jones, Bryan. 2001. *Politics and the Architecture of Choice*. Chicago: University of Chicago Press.

Kaase, Max. 1989. "Mass Participation." In *Continuities in Political Action: A Longitudinal Study of Political Orientations in Three Western Democracies*, M. Kent Jennings et al., eds. Berlin and New York: Walter de Gruyter: 23–66.

Katz, Elihu, and Paul F. Lazarsfeld. 1955. *Personal Influence: The Part Played by People in the Flow of Mass Communications*. Glencoe, Ill.: Free Press.

Katznelson, Ira. 1986. "Working Class Formation: Constructing Cases and Comparisons." In *Working-Class Formation: Nineteenth Century Patterns in Western Europe and the United States*, Ira Katznelson and Aristide Zolberg, eds. Princeton, N.J.: Princeton University Press: 3–44.

References

Kenny, Christopher B. 1994. "The Microenvironment of Attitude Change." *Journal of Politics* 56(3): 715–28.

Key, V. O., Jr. 1961. *Public Opinion and American Democracy*. New York: Knopf.

Key, V. O., Jr., with Milton C. Cummings, Jr. 1966. *The Responsible Electorate: Rationality and Presidential Voting 1936–60*. Cambridge, Mass.: Harvard University Press.

Key, V. O., Jr., and Frank Munger. 1959. "Social Determinism and Electoral Decision: The Case of Indiana." In *American Voting Behavior*, Eugene Burdick and Arthur Brodbeck, eds. Glencoe, Ill.: The Free Press: 281–99.

Kiewiet, Donald R. 1983. *Macro-economics and Micro-politics: The Electoral Effects of Economic Issues*. Chicago: University of Chicago Press.

King-Cassas, Brooks, Damon Tomlin, Cedric Anen, Colin Camerer, Steven R. Quartz,and P. Read Montague. 2005. "Getting to Know You: Reputation and Trust in a Two-Person Economic Exchange." *Science* 308: 78–82.

Kingston, Paul William, and Steven E. Finkel. 1987. "Is There a Marriage Gap in Politics?" *Journal of Marriage and the Family* 49(1): 57–64.

Knack, Stephen. 1992. "Civic Norms, Social Sanctions, and Voter Turnout." *Rationality and Society* 4: 133–56.

1994. "Does Rain Help the Republicans? Theory and Evidence on Turnout and the Vote." *Public Choice* 79(1–2): 187–209.

Knickmeyer, Rebecca, Simon Baron-Cohen, Peter Ragatt, and Kevin Taylor. 2005. "Foetal Testosterone, Social Relationships, and Restricted Interests in Children." *Journal of Child Psychology and Psychiatry* 46(2): 198–210.

Kohler, Ulrich. 2002. *Der demokratische Klassenkampf. Zum Zusammenhang von Sozialstruktur und Parteipräferenz*. Frankfurt a.M and New York: Campus.

2005. "Changing Class Locations and Partisanship in Germany." In *The Social Logic of Politics: Personal Networks as Contexts for Political Behavior*, Alan S. Zuckerman, ed. Philadelphia: Temple University Press: 117–32.

Kotler-Berkowitz, Laurence. 2001. "Religion and Voting Behaviour in Great Britain: A Reassessment." *British Journal of Political Science* 31: 523–54.

2005. "Friends and Politics: Linking Diverse Friendship Networks and Political Participation." In *The Social Logic of Politics: Personal Networks as Contexts for Political Behavior*, Alan S. Zuckerman, ed. Philadelphia: Temple University Press: 152–70.

Kroh, Martin, and Peter Selb. 2005. "Partisanship Inheritance: Blindfolding the Next Generation." Berlin: German Institute for Economic Research.

Kroh, Martin, and M. Spiess, 2004. *Documentation of Sample Sizes and Panel Attrition in the German Socio Economic Panel (SOEP) 1984–2004*. Berlin: DIW Data Documention 6.

Laland, Kevin M. 2001. "Imitation, Social Learning, and Preparedness as Mechanisms of Bounded Rationality." In *Bounded Rationality: The Adaptive Toolbox*, Gerd Gigerenzer and Reinhard Selten, eds. Cambridge, Mass.: MIT Press: 233–48.

Lane, Robert E. 1959. *Political Life: Why People Get Involved in Politics*. Glencoe, Ill.: The Free Press.

Lau, Richard R. 2003. "Models of Decision Making." In *Oxford Handbook of Political Psychology*, David O. Sears, Leonie Huddy, and Robert Jervis, eds. New York: Oxford University Press: 19–59.

References

Lazarsfeld, Paul F., Bernard Berelson, and Hazel Gaudet. 1968 [1948]. *The People's Choice: How the Voter Makes Up His Mind in a Presidential Campaign*, 3rd edition. New York: Columbia University Press.

Lemann, Nicholas. 2003. "The Controller: Karl Rove's Plans to Take Over Politics." *New Yorker* May 12: 68–83.

Levi, Primo. 1988. *The Drowned and the Saved*. London: Abacus, tr. Raymond Rosenthal.

Levine, Jeffrey. 2005. "Choosing Alone? The Social Network Basis of Modern Political Choice." In *The Social Logic of Politics: Personal Networks as Contexts for Political Behavior*, Alan S. Zuckerman, ed. Philadelphia: Temple University Press: 132–51.

Lewin, Kurt. 1948. *Resolving Social Conflicts: Selected Papers in Group Dynamics*. New York: Harper and Brothers.

Lichbach, Mark Irving. 1996. *The Rebel's Dilemma*. Ann Arbor: University of Michigan Press.

1997. *The Cooperator's Dilemma*. Ann Arbor: University of Michigan Press.

Lin, Ann Chih. 2005. "Networks, Gender, and the Use of State Authority: Evidence from a Study of Arab Immigrants in Detroit." In *The Social Logic of Politics*, Alan S. Zuckerman, ed. Philadelphia: Temple University Press: 171–84.

Long, Scott J. 1997. *Regression Models for Categorical and Limited Dependent Variables*. Thousand Oaks, Calif.: Sage Publications.

Lupia, Arthur, and Matthew McCubbins. 2000. "The Institutional Foundations of Political Competence: How Citizens Learn What They Need to Know." In *Elements of Reason: Cognition, Choice, and the Bounds of Rationality*, Arthur Lupia, Matthew D. McCubbins, and Samuel L. Popkin, eds. New York: Cambridge University Press: 47–66.

Macdonald, Stuart Elaine, Olla Listhaug, and George Rabinowitz. 1991. "Issues and Party Support in Multiparty Systems." *American Political Science Review* 85: 1107–31.

Macy, Michael W. 1998. "Social Order and Emergent Rationality." In *What Is Social Theory? The Philosophical Debates*, Alan Sica, ed. Malden, Mass.: Blackwell: 219–37.

Maimonides. 1963. *The Guide of the Perplexed*. Chicago: University of Chicago Press, tr. Shlomo Pines.

1972. *Mishne Torah. The Book of Knowledge. Laws of Beliefs.* 6:1. Jerusalem: Hotzaat Makor (Hebrew).

Manski, Charles. 1993. "Identification of Endogenous Social Effects: The Reflection Problem." *Review of Economic Studies* 60: 532–42.

March, James G. 1953-4. "Husband and Wife Interaction over Political Issues." *Public Opinion Quarterly* 17: 4, 461–70.

March, James G., with Chip Heath. 1994. *A Primer on Decision Making: How Decisions Happen*. New York: Free Press.

Markman, Arthur B., and Douglas L. Medin. 2002. "Decision Making." In *Stevens' Handbook of Experimental Psychology*. New York: Wiley. Available on *http://www.mrw.interscience.wiley.com/shep/articles/paso210/sect1*. Web site posting: July 15, 2002.

Mebane, Walter, Jr. 2004. "Cuing and Coordination in American Politics." Prepared for delivery at 2004 Political Methodology Summer Meeting, Stanford, Calif., July 29–31.

References

Meehl, Paul E. 1977. "The Selfish Voter Paradox and the Thrown-Away Vote Argument." *American Political Science Review* 71: 11–30.
Menand, Louis. 2004. "The Unpolitical Animal: How Political Scientists Understand Voters." *The New Yorker*. August 30: 96.
Merton, Robert K. 1957. *Social Theory and Social Structure*. New York: The Free Press. Revised and enlarged edition.
Miller, Harvey J. 2004. "Tobler's First Law and Spatial Analysis." *Annals of the Association of American Geographers* 94(2): 284–9.
Miller, Robert L., Rick Wilford, and Freda Donoghue. 1999. "Personal Dynamics as Political Participation." *Political Research Quarterly* 52(2): 269–92.
Miller, Warren E., and J. Merrill Shanks. 1996. *The New American Voter*. Cambridge, Mass.: Harvard University Press.
MORI Research Report. 2004. *The Captain's First Mate*. Available on http://www.chumento.co.uk/infospace.
Mosca, Gaetano. 1939. *The Ruling Class: Elementi di Scienza Politics*. New York: McGraw-Hill. Edited and revised by Arthur Livingson and translated by Hannah D. Kahn.
Müller, Wilhelm. 1999. "Class Cleavages in Party Preferences in Germany-Old and New." In *The End of Class Politics? Class Voting in Comparative Context*, Geoffrey Evans, ed. New York: Oxford: 137–80.
Mutz, Diana C. 1998. *Impersonal Influence: How Perceptions of Mass Collectives Affect Political Attitudes*. New York: Cambridge University Press.
Mutz, Diana C., and Jeffrey Mondak. 1997. "Dimensions of Sociotropic Behavior: Group-Based Judgments of Fairness and Well-Being." *American Journal of Political Science* 41(1): 284–308.
Newcomb, Theodore M., Ralph H. Turner, and Philip E. Converse. 1964. *Social Psychology: The Study of Human Interaction*. New York: Holt, Rinehart, and Winston.
Niemi, Richard G., Roman Hedges, and M. Kent Jennings. 1977. "The Similarity of Husbands' and Wives' Political Views." *American Politics Quarterly* 5(2): 133–48.
Niemi, Richard G., and M. Kent Jennings. 1991. "Issues and Inheritance in the Formation of Party Identification." *American Journal of Political Science* 35(4): 970–88.
Norpoth, Helmut. 1984. "The Making of a More Partisan Electorate in Germany." *British Journal of Political Science* 14(1): 53–71.
Nye, Judith L., and Aaron M. Brower. 1996. *What's Social about Social Cognition?: Research on Socially Shared Cognition in Small Groups*. Thousand Oaks, Calif.: Sage Publications.
Olson, Mancur, Jr. 1965. *The Logic of Collective Action*. Cambridge, Mass.: Harvard University Press.
Oygard, Lisbet, Knut-Inge Klepp, Grethe S. Tell, and Odd D. Vellar. 1995. "Parental and Peer Influences on Smoking among Young Adults: Ten-Year Follow-up of the Oslo Youth Study Participants." *Addiction* 90(4): 561–9.
Oyserman, Daphna, and Martin J. Packer. 1996. "Social Cognition and Self-Concept: A Socially Contextualized Model of Identity." In *What's Social About Social Cognition?: Research on Socially Shared Cognition in Small Groups*, Judith L. Nye and Aaron Brower, eds. Thousand Oaks, Calif.: Sage Publications: 175–204.

References

Page, Benjamin I., and Robert Y. Shapiro. 1992. *The Rational Public: Fifty Years of Trends in Americans' Policy Preferences*. Chicago: University of Chicago Press.

Parry, Geraint, George Moyser, and Neil Day. 1992. *Political Participation and Democracy in Britain*. New York: Cambridge University Press.

Pattie, Charles J., and Ron J. Johnston. 1999. "Context, Conversation and Conviction: Social Networks and Voting in the 1992 British General Election." *Political Studies* 47: 877–89.

———. 2000. "People Who Talk Together Vote Together: An Exploration of the Neighbourhood Effect in Great Britain." *Annals, Association of American Geographers* 90: 41–66.

Pattie, Charles J., and Ron J. Johnston. 2001. "Talk as a Political Context: Conversation and Electoral Change in the British Elections, 1992–97." *Electoral Studies* 20(1): 17–40.

Pattie, Charles, Patrick Seyd, and Paul Whiteley. 2004. *Citizenship in Britain: Values, Participation and Democracy*. New York: Cambridge University Press.

Petersson, Olof, Anders Westholm, and Goran Blomberg. 1989. *Medborganas Makt*. Stockholm: Carlssons.

Pienta, Amy M. 2003. "Partners in Marriage: An Analysis of Husbands' and Wives' Retirement Behavior." *The Journal of Applied Gerontology* 22(3): 340–58.

Plassner, Fritz, Christian Scheucher, and Christian Senft. 1999. "Is There a European Style of Political Marketing? A Survey of Political Managers and Consultants." In *Handbook of Political Marketing*, Bruce I. Newman, ed. Thousand Oaks, Calif.: Sage Publications: 90–112.

Powdthavee, Nattavudh. 2005. "For Better and for Worse? A Direct Evidence for Utility Interdependence in Marriage." Department of Economics, University of Warwick, Manuscript.

Rabinowitz, George, and Stuart Elaine Macdonald. 1989. "A Directional Theory of Issue Voting." *American Political Science Review* 83: 93–121.

Richardson, Bradley M. 1991. "European Party Loyalties Revisited." *American Political Science Review* 85: 751–76.

Riesman, David. 1961 [1951]. *The Lonely Crowd: A Study of the Changing American Character*. New Haven, Conn.: Yale University Press.

Riker, William H. 1982. "The Two-Party System and Duverger's Law: An Essay on the History of Political Science." *American Political Science Review* 76: 753–66.

Rilling, James K., David Gutman, Thorsten R. Zeh, Giuseppe Pagnoni, Gregory S. Berns, and Clinton D. Kilts. 2002. "A Neural Basis for Social Cooperation." *Neuron* 35: 395–405.

Robinson, W. S. 1957. "The Statistical Measurement of Agreement." *American Sociological Review* 22(1): 17–25.

Sacerdote, Bruce. 2000. *Peer Effects with Random Assignment: Results for Dartmouth Roommates*. National Bureau of Economic Research Working Paper 7469, Cambridge, Mass. Available on *http://ideas.repec.org/p/nbr/nberwo/7469/html*.

Sanders, David, and Malcolm Brynin. 1999. "The Dynamics of Party Preference Change in Britain," *Political Studies* 47: 219–39.

Sapiro, Virginia. 2004. "Not Your Parents' Political Socialization: Introduction for a New Generation." *Annual Review of Political Science* 7: 1–23.

References

Sargent, James D., and Margaret Dalton. 2001. "Does Parental Disapproval of Smoking Prevent Adolescents From Becoming Established Smokers?" *Pediatrics* 108(6): 1256–62.

Saris, Willem E., and Paul Sniderman, eds. 2004. *Studies in Public Opinion: Attitudes, Nonattitudes, Measurement Error, and Change*. Princeton, N.J.: Princeton University Press.

Särlvick, Bo, and Ivor Crewe. 1983. *Decade of Dealignment: The Conservative Victory of 1979 and Electoral Trends in the 1970s*. New York: Cambridge University Press.

Sartori, Anne E. 2003. "An Estimator for Some Binary-Outcome Selection Models Without Exclusion Restrictions." *Political Analysis* 11: 111–38.

Sartori, Giovanni. 1976. *Parties and Party Systems: A Framework for Analysis*. New York: Cambridge University Press.

Sayer, Liana C., Suzanne M. Bianchi, and John P. Robinson. 2004. "Are Parents Investing Less in Children? Trends in Mothers' and Fathers' Time with Children." *American Journal of Sociology* 110(1): 1–43.

Sayer, Liana C., Anne H. Gauthier, and Frank F. Furstenberg. 2004. "Educational Differences in Parents' Time with Children: Cross-National Variations." *Journal of Marriage and the Family* 66: 1149–66.

Schachter, Stanley, and Michael Gazzaniga, eds. 1989. *Extending Psychological Frontiers: Selected Works of Leon Festinger*. New York: Russell Sage Foundation.

Schickler, Eric, and Donald Philip Green. 1997. "The Stability of Party Identification in Western Democracies." *Comparative Political Studies* 30(4): 450–83.

Schmitt, Herman, and Soren Holmberg. 1995. "Political Parties in Decline?" In *Citizens and the State: A Changing Relationship?* Dieter Fuchs and Hans-Dieter Klingemann, eds. New York: Oxford University Press: 95–133.

Schmitt, Karl. 1998. "The Social Bases of Voting Behavior in Unified Germany." In *Stability and Change in German Elections: How Electorates Merge, Converge, or Collide*, Christopher. J. Anderson and Carsten Zelle, eds. Westport, Conn.: Praeger: 33–54.

Schmitt-Beck, Rüdiger. 2003. "Mass Communication, Personal Communication and Vote Choice: The Filter Hypothesis of Media Influence in Comparative Perspective." *British Journal of Political Science* 33: 233–59.

2006. Rüdiger. "New Modes of Campaigning." In *Oxford Handbook of Political Behavior*, Russell J. Dalton and Hans-Dieter Klingemann, eds. New York: Oxford University Press.

Schmitt-Beck, Rüdiger, and David M. Farrell. 2002. "Studying Political Campaigns and their Effects." In *Do Political Campaigns Matter? Campaign Effects in Elections and Referendums*, David M. Farrell and Rüdiger Schmitt-Beck, eds. London and New York: Routledge: 1–21.

Schmitt-Beck, Rüdiger, Stefan Weick, and Bernhard Christoph. 2006. "Shaky Attachments: Individual-Level Stability and Change of Partisanship among West German Voters, 1984-2001." *European Journal of Political Research* 45: 581–601.

Sears, David O., Leonie Huddy, and Robert Jervis. 2003. "The Psychologies Underlying Political Psychology." In *Oxford Handbook of Political Psychology*, David O. Sears, Leonnie Huddy, and Robert Jervis, eds. New York: Oxford University Press.

References

Selten, Reinhard. 2001. "What Is Bounded Rationality?" In *Bounded Rationality: The Adaptive Toolbox*, Gerd Gigerenzer and Reinhard Selten, eds. Cambridge, Mass.: MIT Press: 13–36.
Shafir, Eldar, and Robyn LeBoeuf. 2002. "Rationality." *American Review of Psychology* 53: 491–517.
Shafir, Eldar, and Amos Tversky. 1995. "Decision Making." In *An Invitation to Cognitive Science, Second Edition* (Volume 3: Thinking), E. E. Smith and D. N. Osherson, eds. Cambridge, Mass.: MIT Press: 77–100.
Shils, Edward A. 1951. "The Study of the Primary Group." In *The Policy Sciences*, Daniel Lerner and Harold D. Lasswell, eds. Stanford, Calif.: Stanford University Press: 44–69.
Shively, W. Phillips. 1977. "Information Costs and the Partisan Life Cycle." 83rd Annual Meeting of the American Political Science Association, Washington, D.C.
Simmel, Georg. 1955. *Conflict and the Web of Group-Affiliations*. New York: The Free Press, tr. Kurt H. Wolf and Reinhard Bendix.
Simon, Herbert A. 1965 [1957]. *Administrative Behavior: A Study of Decision-Making Processes in Administrative Organization*. New York: The Free Press.
    1999. "Comments on Remarks of James Buchanan and Douglass C. North." In *Competition and Cooperation: Conversations with Nobelists on Economics and Politics*, James Alt, Margaret Levi, and Elinor Ostrom, eds. New York: Sage Foundation.
Sinnott, Richard. 1998. "Party Attachment in Europe: Methodological Critique and Substantive Interpretations." *British Journal of Political Science* 28: 627–50.
Smetana, Judith. 2005. "Editor's Notes." *New Directions for Child and Adolescent Development* Summer: 1–3.
Smith, Alison J. 2004. "Who Cares? Fathers and the Time They Spend Looking After Children." Nuffield College, University of Oxford, June 10.
Sniderman, Paul M. 2000. "Taking Sides: a Fixed Choice Theory of Political Reasoning." In *Elements of Reason: Cognition, Choice, and the Bounds of Rationality*, Arthur Lupia, Matthew D. McCubbins, and Samuel L. Popkin, eds. New York: Cambridge University Press: 67–84.
Sniderman, Paul M., Richard A. Brody, and Philip E. Tetlock. 1991. *Reasoning and Choice*. New York: Cambridge University Press.
Sniderman, Paul M., and John Bullock. 2004. "A Consistency Theory of Public Opinion and Political Choice: The Hypothesis of Menu Dependence." In *Studies in Public Opinion: Attitudes, Nonattitudes, Measurement Error and Change*, William E. Sarris and Paul M. Sniderman, eds. Princeton, N.J.: Princeton University Press: 337–58.
Steinberg, Laurence. 2001. "We Know Some Things: Adolescent-Parent Relationships in Retrospect and Prospect." *Journal of Research on Adolescence* 11: 1–19.
    2005. "Psychological Control: Style or Substance." *New Directions for Child and Adolescent Development* Summer: 71–8.
Steinberg, Laurence, and Jennifer S. Silk. 2002. "Parenting Adolescents." In *Handbook of Parenting*, M. H. Bornstein, ed. Mahwah, N.J.: Erlbaum: 103–34.
Stoker, Laura,and M. Kent Jennings. 1995. "Life Cycle Transitions and Political Participation: The Case of Marriage." *The American Political Science Review* 89(2): 421–33.

References

 2005. "Political Similarity and Influence between Husbands and Wives." In *The Social Logic of Politics*, Alan S. Zuckerman, ed. Philadelphia: Temple University Press: 51–74.
Straits, Bruce C. 1990. "The Social-Context of Voter Turnout." *Public Opinion Quarterly* 54(1): 64–73.
 1991. "Bringing Strong Ties Back In: Interpersonal Gateways to Political Information and Influence." *Public Opinion Quarterly* 55(3): 432–48.
Taber, Charles H. 2003. "Information Processing and Public Opinion." In *Oxford Handbook of Political Psychology*, David O. Sears, Leonnie Huddy, and Robert Jervis, eds. New York: Oxford University Press: 433–76.
Tedin, Kent L. 1974. "The Influence of Parents on the Political Attitudes of Adolescents." *American Political Science Review* 68: 1579–92.
Tetlock, Philip E., and Jennifer S. Lerner, 1999. "The Social Contingency Model: Identifying Empirical and Normative Boundary Conditions on the Error-and-Bias Portrait of Human Nature." In *Dual-Process Theories in Social Psychology*, Shelly Chaiken and Yaakov Trope, eds. New York: Guilford Press: 571–85.
Tinsley, Howard E. A., and David J. Weiss. 2000. "Interrater Reliability and Agreement." In *Handbook of Applied Multivariate Statistics and Mathematical Modeling*. New York: Academic Press: 96–125.
Tocqueville, Alexis De. 1969 [1858]. *Democracy in America*. J. P. Mayer, ed. New York: Doubleday-Anchor, tr. George Lawrence.
Ullmann-Margalit, Edna. 2005. "Big Decisions: Opting, Converting, Drifting." *Center for the Study of Rationality*. Hebrew University of Jerusalem. Discussion Paper #409, November.
Ullmann-Margalit, Edna, and Sidney Morgenbesser. 1977. "Picking and Choosing." *Social Research* 44: 757–85.
Van Deth, Jan W. 1989. "Interest in Politics." In *Continuities in Political Action: A Longitudinal Study of Political Orientations in Three Western Democracies*, M. Kent Jennings et al., eds. Berlin and New York: Walter de Gruyter: 275–312.
Ventura, Raphael, 2001. "Family Political Socialization in Multiparty Systems." *Comparative Political Studies* 34(6): 666–91.
Verba, Sidney. 1961. *Small Groups and Political Behavior: A Study of Leadership*. Princeton, N.J.: Princeton University Press.
Verba, Sidney, Kay Schlozman, and Henry Brady. 1995. *Voice and Equality: Civic Voluntarism in American Politics*. Cambridge, Mass.: Harvard University Press.
Verba, Sidney, Kay Schlozman, and Nancy Burns. 2005. "Family Ties: Understanding the Intergenerational Transmission of Political Participation." In *The Social Logic of Politics*, ed. Alan S. Zuckerman. Philadelphia: Temple University Press: 95–116.
Walsh, Katherine Cramer. 2004. *Talking about Politics: Informal Groups and Social Identity in American Life*. Chicago: University of Chicago Press.
Weakliem, David L., and Anthony Heath. 1999. "The Secret Life of Class Voting: Britain, France, and the United States since the 1930s." In *The End of Class Politics? Class Voting in Comparative Context*, Geoffrey Evans, ed. New York: Oxford University Press: 97–133.
Weber, Max. 1978 [1922]. *Economy and Society*. Gunter Roth and Claus Wittich, eds. University of California Press.

References

Westholm, Anders. 1991. *The Political Heritage: Testing Theories of Political Socialization.* Ph.D. dissertation, Uppsala University.
  1999. "The Perceptual Pathway: Tracing the Mechanisms of Political Value Transfer Across Generations." *Political Psychology* 20: 525–51.
Whiteley, Paul, and Patrick Seyd. 1998. "Labour's Grassroots Campaign in 1997." In *British Elections and Elections Review* VIII, David Denver, Justin Fisher, Philip Cowley, and Charles Pattie, eds. London: Cass: 191–207.
  2003. "Party Election Campaigning in Britain: The Labour Party." *Party Politics* 9(5): 637–51.
Wielhouwer, Peter W. 2003. "In Search of Lincoln's Perfect List: Targeting in Grassroots Campaigns." *American Politics Research* 31(6): 632–69.
Wilson, Sven E. 2002. "The Health Capital of Families: An Investigation of the Inter-spousal Correlation of Health Status Correlation." *Social Science and Medicine* 55: 1157–72.
Winkelmann, Rainer. 2000. *Econometric Analysis of Count Data.* New York: Springer.
Wooldridge, Jeffrey M. 2003. *Introductory Econometrics: A Modern Approach.* Cincinatti, Oh.: South-Western College Publishing.
Zelle, Carsten. 1998. "A Third Face of Dealignment? An Update on Party Identification in Germany 1971–94." In *Stability and Change in German Elections: How Electorates Merge, Converge or Collide,* Christopher J. Anderson and Carsten Zelle, eds. Westport, Conn.: Praeger Publishers.
Zuckerman, Alan S. 1991. *Doing Political Science: An Introduction to Political Analysis.* Boulder, Colo.: Westview Press.
  ed. 2005a. *The Social Logic of Politics.* Philadelphia: Temple University Press.
  2005b. "Returning to the Social Logic of Politics." In *The Social Logic of Politics,* Alan S. Zuckerman, ed. Philadelphia: Temple University Press: 3–20.
Zuckerman, Alan S., Jennifer Fitzgerald and Josip Davić. 2005. "Do Couples Support the Same Political Parties? Sometimes: Evidence from British and German Household Panel Surveys." In *The Social Logic of Politics,* Alan S. Zuckerman, ed. Philadelphia: Temple University Press: 75–94.
Zuckerman, Alan S., and Laurence A. Kotler-Berkowitz. 1998. "Politics and Society: Political Diversity and Uniformity in Households as a Theoretical Puzzle." *Comparative Political Studies* 31(4): 464–97.
Zuckerman, Alan S., Laurence A. Kotler-Berkowitz, and Lucas A Swaine. 1998. "Anchoring Political Preferences: The Importance of Social and Political Contexts and Networks in Britain." *European Journal of Political Research* 33(3): 285–321.
Zuckerman, Alan S., and Martin Kroh. 2006. "The Social Logic of Bounded Partisanship in Germany: A Comparison of West Germans, East Germans, and Immigrants." *Comparative European Politics* 4(1): 65–93.
Zuckerman, Alan S., Nicholas A. Valentino, and Ezra W. Zuckerman. 1994. "A Structural Theory of Vote Choice: Social and Political Networks and Electoral Flows in Britain and the United States." *Journal of Politics* 56(4): 1008–33.

Index

Achen, Christopher, xvi, 6–7, 29–30, 33, 46, 92
Acock, Alan, 110
Administrative Behavior (Simon), 19
Age cohorts (*see also* partisan constancy, partisan preference, partisan support, turnout)
    Defined, 126
    Indicators in BHPS, 57, 62
    Indicators in GSOEP, 53, 56
Agent-based models, 3
Aggregate (macro) analysis of partisanship, xvii, xxvi, 16, 24, 27, 34
Alford, C. Fred, 3, 29
Allen, Woody, 21
American National Election Surveys (ANES), 12, 20
American Voter, The (Campbell, Converse, Miller, Stokes), 12, 14, 17, 19, 22–5, 27
Andersen, Robert, xxvii
Anderson, Christopher, xxvii, 28, 56, 125
Anen, Cedric, xv
Aquinas, Thomas, 10–11
Aristotle, 1, 9–10, 71, 80, 159
Aron, Art, 28
Aron, Elaine, 28
Arrow, Kenneth, 4, 29
Associative (or assortative) mating, 69, 71, 73, 80

Aumann, Robert, xvii, 7, 148
Axelrod, Robert, 2–3, 29

Baker, Kendall, 91
Bannon, Declan, 32, 150–1
Bantle, Christian, 92
Barbee, Anita, 2, 29
Barker, Ernest, 1, 9
Baron-Cohen, Simon, xv
Bartle, John, xxv
Bean, Clive, 72
Beck, Paul Allen, xvi, 28, 70, 91, 95, 127, 129, 153
Becker, Gary, 69
Behavior Revolution, in political science, 12–15, 17, 28, 148, 151, 158
Bengtson, Vern, 110
Berelson, Bernard, xv, 13–16, 150
Bergmann, Knut, 150
Berns, Gregory, xv, 3
Beyer, Mary Alice, 85
BHPS, *see* British Household Panel Survey
Bianchi, Suzanne, 144
Bible, The (*see also* Genesis, Proverbs, Psalms), 9–11, 29
Bion, W. R., 3, 29
Birds of a Feather Flock Together, principle (*see also* associative mating), 10, 80
Blair, Tony, xxvi, 125

Index

Blais, André, xvii, xxv
Blomberg, Goran, 81, 126
Bolster, Anne, xxvii, 56, 125
Bottomore, T. B., 11
Boudon, Raymond, 4, 8
Bounded partisanship, xviii, xxvii, 33, 40, 45–7, 68, 72–3, 83, 91, 98, 102, 122–3, 139, 142, 145–55, 157, 159
Bounded rationality, xvii, 1, 4–5, 27, 29–30
Brady, Henry, 38
British Election Studies (BES), xxv, 20
British elections, 150–1
  1992, 37, 97, 126
  1997, 37, 97, 125–6, 131, 153
  2001, 97, 100, 125–6, 131
British Household Panel Survey (BHPS)
  Descriptive information, xx, xxiv–xxvii
  Limitations of, xx
  Strengths of, xx, xxv
Brody, Richard, 8
Brower, Aaron, 29
Brynin, Malcolm, 56, 124
Burgess, Simon, xxvii, 56, 125
Bürklin, Wilhelm, 47
Burns, Nancy, 28, 72
Butler, David, 124

Caballero, C., xxii, 47
Camerer, Colin, xv
Campbell, Angus, 12–14, 17, 22–4, 27, 29–30, 33, 159
Carlo Cattaneo Institute, 155
Cartwright, Dorwin, 17
Caughlin, John, 89
Children (*see also* fathers, mothers)
  And partisan constancy, 98
  And partisan preference (choice), 105–9
  And party support, 91, 95–7, 100, 102, 105–6, 108, 120
  And political interest, 105–107, 110, 138, 145
  And vote for Conservatives, 135–6
  And vote for Labour, 131–4
  Defined, 92, 94, 111, 130
  Influence on father, 120–1, 131–8
  Influence on mother, 119–21, 123, 131–8
Christian Democratic Union/Christian Social Union (CDU/CSU), Germany
  Constancy of preference (choice) for, 43, 65, 98
  Preference (choice) for, 45, 55–6, 75, 116
  Preference agreement among couples, 80, 84–5
  Preference agreement among family members, 100–1, 119
Christoph, Bernhard, xxi, 43
Church of England, Anglican (*see also* Religious Identity), 74
Cialdini, Robert, 2
Cognitive dissonance (*see also* Festinger, Leon), 17, 18
Coile, Courtney, 89
Collins, W. Andrew, 92
Coltrane, Scott, 144
Columbia School of Electoral Sociology (see also Partisanship, theories of), xv, 13–14, 17, 31, 127
Columnar report, 41, 44–5, 82
Commentary on the Nicomechean Ethics (Aquinas), 11
Conover, Pamela, 71
Conservative Party, Britain
  Constancy of preference (choice) for, 43, 65, 99
  Preference (choice) for, 44, 57, 61, 77, 106, 116
  Preference agreement among couples, 80, 86–7
  Preference agreement among family members, 101
  Voting for, 127, 135, 136
Converse, Philip, 22, 24, 30, 36, 46, 96
Cooke, Lynn, 144, 154

Index

Couples (*see also* households, husbands, wives)
And partisan agreement, 80–1, 83–5, 87
And religion, 81, 83, 86
And social class, 81, 83, 86
Reciprocal political influence, 76–7, 110, 116, 119
Crewe, Ivor, 71, 146
Cross-Partisanship (*see also* partisan switching), 91
Cunningham, Michael, 2, 28

Dalton, Madeline, 92
Dalton, Russell, 35–6, 47, 98
Dasović, Josip, xvi, 3, 71
Davies, James, xvi
Day, Neil, 38
De Graaf, Nan Dirk, xvi, 71
Decision Theory, 27–8
Demonstrations, political, 40
Dennis, Jack, 91
Denver, David, 150
Determinism, social, 14–15, 20, 90
Deutsche Institut für Wirtschaftsforschung (DIW Berlin), xxi
Diamond, Gregory, 146
Discussion, political, xxiii, 8, 12–13, 28, 71–72, 89, 139, 142, 144, 151
In households, 72, 142
In social networks, 3, 20, 27–8, 30, 129
Spouses as discussion partners, 139
Dogan, Mattei, xvi, 71
Donoghue, Freda, 70
Douglas, Mary, 1–3, 29
Downs, Anthony, 8, 20, 25–7, 29–30, 146, 148
Drowned and the Saved, The (Levi), xv
Duck, Steve, 29
Durkheim, Émile, 11, 17
Duverger, Maurice, 151
Dyadic relations, xvi, xxvii, 21, 27, 88, 94, 116, 130–1, 143

East Germany, former German Democratic Republic (GDR), xxi, xxiv, xxviii, 152
Easton, David, 91
Economic concerns (problems), 52, 54
And Partisan Preference (choice), 56
Indicator in BHPS, 57
Indicator in GSOEP, 51
Economic Theory of Democracy, An (Downs), 26, 31
Economy and Society (Weber), xv
Education level, xxiii, xxv, 24, 50, 57, 126, 140, 161
And partisan preference (choice), 56, 57
And partisan support, 55, 57
Indicators in GSOEP, 50
Electoral campaigns (mobilization), 125, 148, 150, 151
Calculations of party leaders, 35, 142–3, 148–9, 151, 153–4, 159
Segmenting the political market, 154
Target groups, 150
Electoral choice (voting), xvii, xix–xxii, xxv, xxvii, xxviii, 5, 7–8, 12, 14–16, 19–22, 26–27, 29, 31, 45, 71, 122–8, 130–3, 135–6, 138–43, 146, 148–52, 154–8
In Britain, 124
Studies of, 12
Electoral Sociology (*see also* Columbia School of Electoral Sociology), xv, 14
Elmira, New York, Election Study (1940), 14, 16, 22, 28
Elster, Jan, 5
Endogeneity, xxvii, 50, 74
Fallacy of, 49
Erie County, Ohio, election study (1948), 14, 22, 28
Erikson, Robert, xxvii, 34, 56, 125
Eulau, Heinz, 13–14, 21, 159

Index

European Commission, 144
European Union (EU), xxiv, 152
Evans, Geoffrey, xxiii, xxvii, 47
Events, exogenous
   Euro replaces mark, xxiv, 153
   German reunification, xviii, xxi, xxiv, xxvi, 36, 38, 40, 153
   Government change, Britain, xviii, xxvi
   Government change, Germany, xxiv

Falter, J., xxii, 47
Families, *see* couples, households
Farrell, David, 150
Fathers, political influence on children, 91, 111, 119–21, 123, 130–1, 138
Favreault, Melissa, 89
Festinger, Leon, 13, 17–18, 22, 25
Field theory, 17
Finkel, Steven, xvi, 150
Fiorina, Morris, 30, 33
Fisher, Justin, 150
Fisher, Kimberly, 144
Fitzgerald, Jennifer, xvi, 3, 71
Forza Italia (Italy), 155
Fowler, James, 8, 29, 31, 127
Franklin, Mark, 47
Free Democratic Party (FDP), Germany, xxii, 45
Freud, Sigmund, 3, 29
Friends and political discussion, 8–11, 16, 30, 38, 72, 89, 123, 144
Funnel of causality (*see also American Voter, The*), 17, 22
Furstenberg, Frank, Jr., 144

Gaudet, Hazel, xv, 14, 15
Gauthier, Anne, 144
Gazzaniga, Michael, 18
Gender, *see* husbands, wives
Genesis, book of (*The Bible*), 1, 71
German elections, 36, 96, 150–1
German Socio-economic Panel Study (GSOEP)
   Descriptive information, xx, xxi–xxiv
   Limitations of, xx
   Strengths of, xx, xxiii, xxvi
Gershuny, Jonathan, 144
Gigerenzer, Gerd, 4, 29
Gimpel, James, 28, 30
Glaser, William, xvi, 71–2
Gluchowski, P. M., 47
Goldthorpe, John, xxiii, 47, 128
Goldthorpe measures, *see* occupation
Gonzalez, Richard, 21
Granovetter, Mark, 29
Gray, Mark, 126
Greek wisdom, 9, 11, 29
Green, Donald, xxii, 42
Green parties, xxii, 45, 95
Greene, Steven, 129
Greene, W. H., 164
Greenstein, Fred, 91
Griffin, Dale, 21
GSOEP, *see* German Socio-economic Panel Study
Guide of the Perplexed, The (Maimonides), 10
Gutman, David, 3

Haider, Lailer, 154
Haisken-DeNew, John, 92
Hands, Gordon, 150
Happiness, *see* well-being
Haslam, S. Alexander, 29
Hays, Bernadette, 72
Health (*see also* well-being), xvi, xx, 89, 124, 127
Heath, Anthony, xvi, 47, 71
Heath, Chip, 29
Heckman Probit Selection model, xix, xxviii, 48, 50–2, 54, 56, 58–60, 63, 65, 69, 101, 103–4, 107–8, 124–5, 128–9, 163, 165
   Described, 48, 163
Heckman, James, 111, 164–5
Heineman, Elizabeth, 154
Hendrick, Clyde, 29
Hendrick, Susan, 29
Henig, Simon, 150
Hernes, Gunnar, 5

Index

Hess, Robert, 91
Heuristics, 4–5, 8, 29, 123
Hobbes, Thomas, 16
Holbrook, Thomas, 150
Holt, Robert, 150
Holzhacker, Ronald, 153
Homer, 10
Homophily, marital, *see* associative mating
Homo politicus, 26
Household income, 38, 50, 52–5, 58–61, 101, 103–4, 107–8
Households
   And partisan agreement, 80–1, 84–6, 92
   And reciprocal influence, 76–7, 116, 119–20, 131, 138
   And religion, 86, 92
   And social class, 86
   Defined, xvi, xx, 111
   Heads of, xxi, 46, 49, 57–69, 110, 123, 146, 157
   Samples, xxi, 94, 97, 111, 130
Houts, Robert, 89
Huckfeldt, Robert, xv, 3, 8, 28–9, 31, 70, 71, 81, 85
Huddy, Leonie, 25
Husbands (*see also* couples, fathers, households, wives)
   Influence on wives, 73–80, 110–122 130–132, 135–36 138
Huston, Ted, 89
Hyman, Herbert, 91

Ickes, William, 29
Iliad, The (Homer), 10, 71
Immigrants, xxi, 152–3
Institute for Social and Economic Research (University of Essex), xxiv, 161
Instrumental Variable Probit Model, xix, 74, 130, 163
Instrumental variables (*see also* Linear Probability Model), xix, 28, 49, 74, 116, 124, 130, 131, 138, 163–4

Interdependence
   Of group members, 17, 23
   Of observations in data, xx
Interpersonal interaction, 13, 23
   Frequency of, xxiii, 1, 6–8, 12, 93, 138, 143, 145, 159
Israel, 101
Italy, 143, 155
Iversen, Torben, 84

Jaros, Dean, 91
Jennings, M. Kent, xvi, 3, 28, 30, 70, 72, 85, 91, 95
Jervis, Robert, 25
Joesch, Jutta, 144
Johnson, Paul, 3, 28–9
Johnson, Richard, 89
Johnston, Ron, xxvii, 20, 28, 30, 56, 71, 125, 128, 141, 154
Jones, Bryan, 3, 4, 29
Jones, Kelvyn, xxvii, 56, 125

Kaase, Max, 38
Kaplan, Norman, 23
Katz, Elihu, 16, 26–8, 71
Katznelson, Ira, 11
Kenny, Christopher, xv, 70
Key, V. O., 13–14, 17, 20, 22, 25, 27, 29, 148, 159
Kiewiet, Donald, 30, 33
Kilts, Clinton, 3
King-Cassas, Brooks, xv, 3
Kingston, Paul William, xvi
Knack, Stephen, 127
Knickmeyer, Rebecca, xv
Kohl, Helmut, xxiv
Kohler, Ulrich, 36, 45, 47
Kotler-Berkowitz, Laurence, xvi, 3, 20, 28, 30, 47, 70–2, 141
Kroh, Martin, xxi, 95, 152

Labour Party, Britain
   Constancy of preference (choice) for, 43, 65, 99
   Preference (choice) for, 44, 59, 77, 106, 110, 116

Index

Labour Party, Britain (*cont.*)
  Preference agreement among couples, 80, 86–7
  Preference agreement among family members, 101, 119
  Voting for, 127, 131, 133
Laland, Kevin, 5
Lane, Robert, 12
Lau, Richard, 25
Laursen, Brett, 92
Lay, J. Celeste, 28, 30
Lazarsfeld, Paul, xv, 13–17, 20, 24–5, 27–8, 31, 150
Learning, social, xvii, 3–6, 16
LeBeouf, Robyn, 29
Lemann, Nicholas, 150
Lerner, Daniel, 5
Lerner, Jennifer, 5
Levi, Primo, xv
Leviathan (Hobbes), 16
Levine, Jeffrey, 28, 30
Lewin, Kurt, 13, 17–18, 23, 25, 27
Liberal Democratic Party, Britain, xxii, xxv, 95, 100, 141, 148
Lichbach, Mark, 3
Life cycle effects (*see also* age cohorts), 50
Life-space (*see also* Lewin), 18
Like to like, principle (*see also* associative mating), 2, 10, 73, 80
Lin, Ann Chih, 30
Linear Probability Model
  Described, xix
  Instrumental variables, 49, 116, 124, 131
Listhaug, Olla, 146
Local Participation Study (UK), 72, 139, 140
Local politics
  Interest in, 38, 139–40
  Participation in, 38, 139, 152
Logic, social, *see* bounded partisanship
Logistic Regression Model (Logit Model), 48, 74, 84, 130, 139, 165
  Ordered Logit Model, 139

Lonely Crowd, The (Riesman), 17
Long, Scott J., 164
Lupia, Arthur, 8

MacAllister, Ian, 150
Macdonald, Stuart Elaine, 146
Mackie, Tom, 47
MacKuen, Michael, 34
Macurdy, Thomas, 111, 165
Macy, Michael, 2
Maimonides, 10
Major, John, 125
Manski, Charles, 49
March, James, xvi, 29, 71–2
Marketing, consumer or electoral, 32, 92, 150
Markman, Arthur, 32
Marx, Karl, 1, 11, 17
McCarty, Craig, 29
McClurg, Scott, 150
McCubbins, Matthew, 8
McPhee, William, xv, 14–16, 150
Mebane, Walter, 29, 31
Mechanisms, social, xviii, 2, 4–5, 12, 15, 18–19, 21, 29, 142, 158
Media, the, 8, 142, 149
Medin, Douglas, 32
Meehl, Paul, 2, 148
Menand, Louis, 123–4
Mendes, Silvia, xxvii, 56, 125
Merton, Robert K., 13, 16–17, 23, 25, 71
Michigan School of Electoral Studies (*see also* partisanship, theories of), 12–13, 17–25, 27–30, 33, 46
Middle class (*see also* social class), 51, 74, 87
  Subjective identity, 127
Miller, Harvey, 2
Miller, Robert, 70
Miller, Warren, 13, 14, 22–4, 27, 29–30, 33, 159
Mimetism, 11
Mishneh Torah, 10
Mondak, Jeffrey, 127
Montague, P. Read, xv

Index

Morganbesser, Sidney, xvii
MORI Research Report, 89
Mosca, Gaetano, 11, 17
Mothers, political influence on children, 91, 111, 119–21, 130–38, 143–45
Moyser, George, 38
Müller, Wilhelm, 47
Mutz, Diana, 8, 127

National elections (*see also* German elections, British elections), xviii, xxii, xxiv, xxvi, 36, 50–1, 55, 57, 96, 126, 130–1, 151
National Opinion Research Center (NORC), 22
Negative partisanship (*see also* partisanship), 146
Neighborhood
  As social context, 27–8, 30
  Attachment to, 140
  Partisan support in, 50, 83
Neighbors
  Frequency of interaction with,
  Political discussion with, 71, 72, 139, 144
  Political preferences of, xv, 10, 14, 34, 83, 123
Newcomb, Theodore, 22
Newlywed couples (*see also* couples), 83, 85
Ney, Steven, 1–3, 29
Nicomechean Ethics (Aristotle), 10, 71
Niemi, Richard, xvi, 28, 91, 95
No party, support for (*see also* partisan support), xix, 6, 30, 32–3, 37, 40–5, 47–8, 73, 80–1, 83, 91, 95, 98–101, 103–6, 120–1, 124, 146, 147, 149, 152–5, 157, 163
Norms, social, 2–4, 13, 25
Norpoth, Helmut, xxii, 47, 50
Northern Ireland, 70, 144
Nye, Judith, 29

Occupation, xxiii, 24, 74–6
And partisan preference (choice), 56, 76
And partisan support, 55
Goldthorpe measures, xxiii, xxv, 47, 52, 54, 58, 60, 74, 128, 161
Indicator in BHPS, xxv, 74
Indicator in GSOEP, xxiii, 74, 116
Olson, Mancur, Jr., 3
Onkeles, Targum, 1, 159
Opinion diversity in social networks (*see also* couples, households), 29
Opposites attract, principle, 2, 10
Ordinary Least Squares (OLS) Regression Model, 62, 111, 130
Oygard, Lisbet,
Oyserman, Daphna, 29

Packer, Martin, 29
Page, Benjamin, 34
Pagnoni, Giuseppe, 3
Palmquist, Bradley, xxii, 24, 30, 33, 41, 42
Pander, Rohini, 154
Parents, political influence on children (*see also* mothers, fathers), 101–2, 105–6, 108, 131, 133, 135
Parry, Geraint, 38
Partisan constancy (stability of partisan preference), xviii, xix, xxi, xxvi–xxviii, 32, 33, 36, 40–6, 48, 49, 62–8, 72, 73, 116, 123–4, 127–38, 141, 157, 158
  And vote choice, 46
  Defined, 32, 124
Partisan preference (choice), xvi, xxiii, xxv, xxviii, 34–7, 45, 47–68, 72, 80, 89, 94–122, 123, 143, 145, 148, 152, 156, 163
  Defined, xvii, 32
  Indicator in BHPS, xxv
  Indicator in GSOEP, xxii
Partisan support (for any party)
  Defined, xvi, 32
  Indicator in GSOEP, xxii
  Indicator in BHPS, xxv, 156 (*see also* no party)

Index

Partisan switching (*see also* partisan constancy), xviii, xxvi, xxvii, 42, 44–5, 48, 98, 150
Partisanship
  Aggregate level, 44–5, 100
  National, 51, 126, 131
  Regional, xxiv, xxv, 51, 57, 63, 65, 83, 85, 90, 102, 110, 126, 127, 131
  Secular decline of, xviii, xxvi, 55, 59, 97, 102, 153
  Theories of, xv, xviii, 9, 12–28, 30–1, 33, 46, 127
    (*see also* partisan constancy, partisan preference, partisan support, bounded partisanship)
Party leaders, calculations of, 35, 142–3, 148–9, 151, 153–4, 159
Party of Democratic Socialism (PDS), Germany, xxii, xxviii, 152
Paskeviciute, Aida, 28
Pattie, Charles, 20, 28, 30, 38, 71, 128, 141, 154
People's Choice, How the Voter Makes Up His Mind in a Presidential Campaign, The (Lazarsfeld, Berelson, Gaudet), 27
People's Choice, The (Key and Munger), 14–15, 25
Personal Influence (Katz, Lazarsfeld), 16, 26–27, 71
Petersson, Olof, 81, 126
Pienta, Amy, 89
Plassner, Fritz, 150
Poisson Model, 164
Political interest, xvi, xxiv, xxv, xxvii, 8, 35, 38–40, 46, 50, 53, 57, 68, 73–9, 83–5, 87, 89–92, 101–2, 105, 107, 112–18, 124, 126, 131, 145, 149, 158, 161–2
  And partisan agreement, 73, 80, 83, 85
  And partisan constancy, 9, 65
  And partisan preference (choice), 9, 62, 72, 74, 77–8, 89, 116
  And partisan support, 9, 53, 74, 107, 110
  And political influence, xvi, xxiii, 91, 116, 120, 138, 145
  And turnout, 124
Political parties, *see* CDU/CSU, Conservative Party, Labour Party, SPD
Politics, The (Aristotle), 1
Polychoric correlation, 41, 116, 131
Powdthavee, Nattavudh, xvi, 89
Primary groups, *see* couples, households
Propinquity, principle, 2, 141
Propper, Carol, xxvii, 56, 125
Proverbs, book of (*The Bible*), 10
Psalms, book of (*The Bible*), 10
Public Opinion and American Democracy (Key), 26

Quartz, Steven, xv

Rabinowitz, George, 146
Ragatt, Peter, xv
Rallings, Collin, 72, 139
Rational choice theory, xvii, 3, 6, 20, 26, 27, 29–30, 33, 148
  And partisanship (*see also* partisanship, theories of), 20
  Limitations of, 19, 29
    (*see also* bounded rationality, social logic of politics)
Reciprocal political influence, *see* couples, households
Reference Group Theory and Voting Behavior (Kaplan), 23
Relationships, analytical (*see also* determinism)
  Determined vs. probabilistic, xvi, 2, 31, 69, 156
Relationships, social
  As social contexts (*see also* discussion, social circles), xv, xvi, xxi, 7, 138
  Dyadic, xvi, 21, 27, 74, 81, 83, 85, 88, 94, 116, 143

Index

Religious attendance, 51, 75–6, 78–9, 84–5, 90, 92
  And partisan agreement, 84–8
  And partisan constancy, 63–6
  And partisan preference (choice), 51–5, 57–61, 64–7, 74–80, 110
  And partisan support, 51–5, 57–61
  Indicator in BHPS, xxv, 65, 74, 162
  Indicator in GSOEP, xxiii, 65, 74, 162
Religious identification, 51–55, 58–61, 63, 73–6, 78–9, 81, 84–7, 92–3, 102, 104, 107–8, 116, 126–9, 131, 161–2
  And partisan agreement, 80, 83, 84–8
  And partisan constancy, 63–6
  And partisan preference (choice), 51–5, 57–61, 64–7, 74–80, 110
  And partisan support, 51–5, 57–61, 110
  Indicator in BHPS, xxv, 51, 65, 74, 162
  Indicator in GSOEP, xxiii, 65, 74
Responsible Electorate, The (Key), 25
Rho statistic, 53, 55, 59, 61, 103–4, 107–8, 127–9, 163
Richardson, Bradley, 47
Riesman, David, 17
Riker, William, 2, 148
Rilling, James, 3
Robinson, John, 144
Robinson, W. S., 41
Rosenbluth, Frances, 84
Rove, Karl, 150
Rubel, Maximilien, 11

Sacerdote, Bruce, 49
Sanders, David, 56, 124
Sapiro, Virginia, 91
Sargent, James, 92
Saris, Willem, 46
Sarker, Rebecca, xxvii, 56, 125
Särlvick, Bo, 146
Sartori, Anne, 50
Sartori, Giovanni, 35
Sayer, Liana, 144

Scandinavia, 155
Schachter, Stanley, 18
Scheucher, Christian, 150
Schickler, Eric, xxii, 24, 30, 33, 41, 42
Schlozman, Kay, 28, 38, 72
Schmitt, H., 36, 47
Schmitt, K., 47
Schmitt-Beck, Rüdiger, xxi, 8, 36, 43, 70, 150, 153
Schoen, H., xxii, 47
Schrott, Peter, 150
Scottish National Party (SNP), xxii, xxv, 95
Searing, Donald, 71
Sears, David O., 25
Selb, Peter, 95
Selection bias, 48, 163
Selten, Reinhard, 4, 29
Senft, Christian, 150
Seyd, Patrick, 38, 150
Shafir, Eldar, 29
Shanks, J. Merrill, 24, 30, 33
Shapiro, Robert, 34
Shils, Edward, 13, 17
Shively, W. Phillips, 30, 33
Silk, Jennifer, 92, 110
Simmel, Georg, 3, 11, 16–18
Simon, Herbert, 4, 18–19, 29, 158
Sinnott, Richard, 47
Small world studies (*see also* Fowler), 29
Smeeding, Timothy, 144
Smetana, Judith, 94, 110
Smith, Alison, 144
Sniderman, Paul, 8, 46, 151
Snyder, James, 111, 165
Social Capital Benchmark Study USA, (2000), 28
Social circles
  As social contexts, 19–20, 22, 25–6, 31, 71–2, 139, 144
  Immediate (intimate), xxv–xxvii, 8, 12, 17, 19–20, 22, 25–6, 31, 56, 69, 71, 78, 80, 89, 124–5, 127, 135, 139, 141, 142, 146, 151–4, 159

Index

Social class, subjective identification
  Agreement among family members, 92–3
  Agreement between household partners, 81, 83
  And partisan constancy, 66–8
  And partisan preference (choice), 57–62, 84–7, 106, 110
  And partisan support, 57–61
  And vote choice, Britain, 127–9, 131–5
  Indicator in BHPS, 127 (*see also* occupation)
Social Democratic Party (SPD), Germany
  Partisan constancy, 64–5, 67, 98–9
  Preference (choice) for, 44, 51–3, 55–6, 75–7, 102–3, 105–6, 116, 121
  Preference agreement among couples, 80–5
  Preference agreement among family members, 101, 112–13, 116, 119
  Social intimates, *see* couples, households
Social logic of bounded partisanship, xviii, xix, xxiii, xxvii, xxviii, 9, 27, 30, 33, 91, 123, 139, 141, 147, 151–4, 159
Social logic of politics, xviii–xx, xxviii, 9, 11–21, 23, 25–8, 30, 46, 48, 138, 142
  And rational choice theory, 29, 30
Social networks, xvi, 13, 29, 85, 127
Social organizations, membership in, 6, 38, 126
Social ties, 6
Sometimes true theories, *see* mechanisms
Spiess, Katharina, 144
Spiess, M., xxi
Spouses, *see* couples, households, husbands, wives
Sprague, John, xv, 3, 8, 28–9, 31, 70, 71, 81, 85
Steinberg, Laurence, 92, 110

Stimson, James, 34
Stoker, Laura, xvi, 3, 28, 30, 70, 72, 85
Stokes, Donald, 12–14, 22–4, 27, 29, 124, 159
Straits, Bruce, xv, 127
Swaine, Lucas, 20, 71, 72
Sweden, 81, 101, 126, 155

Taber, Charles, 34
Taylor, Kevin, xv
Tedin, Kent, xvi, 91
Tetlock, Philip, 5, 8
Thrasher, Michael, 72, 139
Time-series data, xx, 84
Tinsley, Howard, 41
Tobler, 2
Tocqueville, Alexis de, 11, 17
Tomlin, Damon, xv
Torney, Judith, 91
Trade union membership, 5, 51, 63, 74–6, 78–9, 84–5, 87, 90, 101–2, 106, 112–18, 127–8, 131–2, 136, 162–3
  And partisan agreement, 83–8
  And partisan constancy, 63–6
  And partisan preference (choice), 5, 51–5, 57–61, 64–7, 74–80
  And partisan support, 51–5, 57–61
  And vote choice, Britain, 127–132
  Indicator in BHPS, xxv, 75, 162
  Indicator in GSOEP, xxiii, 50, 51, 75, 162
Treatise on Happiness (Aquinas), 11
Trust, social, xxiii, 2, 5, 8, 25, 89, 91, 158, 159
  And social learning, 5, 8, 15, 157, 159
  In households, xxiii, 7, 8, 89, 91, 93, 111, 143, 145, 158, 159
Turner, John, 29, 150
Turner, Ralph, 22
Turnout, electoral, xix, xxv–xxviii, 8, 29, 124–7, 140–1, 150–1, 156–8
  Effects of living alone on, xvi, 51, 83, 126, 135, 140–1
Tverdova, Yuliya, xxvii, 56, 125

Index

Tversky, Amos, 29
Two-Stage Instrumental Probit Model, 111, 165–6

Ullmann-Margalit, Edna, xvii
United States, xxii, xxv, 70, 144, 150, 155
Utility maximization, *see* rational choice theory

Valen, Henry, 47
Van Deth, Jan, 38
Ventura, Raphael, xvi, 95, 101
Verba, Sidney, 13, 17, 28, 38, 72
Voluntary activities
    Indicator in BHPS, 163
    Indicator in GSOEP, xxiii, 50, 162
Von Wilamowitz-Moellendorff, 47
Voting, *see* turnout, electoral choice

Walsh, Katherine Cramer, 71
Weakliem, David, 47
Weber, Max, xv, 11, 16–18
Weick, Stefan, 36, 43
Weiss, David, 41
Well-being, subjective (*see also* turnout), 126, 128, 129
Welsh Nationalist Party (Plaid Cymru), Britain, xxii, xxv, 95
Westholm, Anders, 81, 95, 101, 126
Whitehurst, Robert, 85

Whiteley, Paul, 150
Wickert, Wolfram, 150
Wielhouwer, Peter, 150
Wilson, Sven, xvi
Winkelmann, Rainer, 63, 164
Wives (*see also* couples, households, husbands, mothers)
    Centrality in families, 116, 119, 120–1, 130, 135, 138, 143–5, 153
    Influence on husbands, 73–80, 110–22, 130–2, 135, 137
Wooldridge, Jeffrey, 48, 50, 164
Working class (*see also* occupation, social class), 74, 86–7, 162
    Subjective identity, 127
Workmates and political discussion, 8, 14, 71, 72, 144
World Values Survey (WVS), 38–9

Young person, *see* children

Zeh, Thorsten, 3
Zelle, Carsten, xxii, 36
Zero-Inflated Negative Binomial model (ZINB), 63, 65, 163–4
    And partisan constancy, 62
    Described, 164
Zuckerman, Alan S., xv, xvi, xxi, 3, 9, 11, 20, 28, 30, 70–2, 141, 152, 153
Zuckerman, Ezra, 71

Books in the Series

Asher Arian, *Security Threatened: Surveying Israeli Opinion on Peace and War*

James DeNardo, *The Amateur Strategist: Intuitive Deterrence Theories and the Politics of the Nuclear Arms Race*

Robert S. Erikson, Michael B. MacKuen, and James A. Stimson, *The Macro Polity*

James L. Gibson and Amanda Gouws, *Overcoming Intolerance in South Africa: Experiments in Democratic Persuasion*

John R. Hibbing and Elizabeth Theiss-Morse, *Congress As Public Enemy: Public Attitudes Toward American Political Institutions*

John R. Hibbing and Elizabeth Theiss-Morse, *Stealth Democracy: Americans' Beliefs about How Government Should Work*

John R. Hibbing and Elizabeth Theiss-Morse, *What Is It about Government That Americans Dislike?*

Robert Huckfeldt, Paul E. Johnson, and John Sprague, *Political Disagreement: The Survival of Diverse Opinions within Communication Networks*

Robert Huckfeldt and John Sprague, *Citizens, Politics, and Social Communication*

James H. Kuklinski, *Thinking about Political Psychology*

Richard R. Lau and David P. Redlawsk, *How Voters Decide: Information Processing in Election Campaigns*

Arthur Lupia, Mathew McCubbins, and Samuel Popkin, *Elements of Reason: Cognition, Choice, and the Bounds of Rationality*

George E. Marcus, John L. Sullivan, Elizabeth Theiss-Morse, and Sandra L. Wood, *With Malice Toward Some: How People Make Civil Liberties Judgments*

Diana C. Mutz, *Impersonal Influence: How Perceptions of Mass Collectives Affect Political Attitudes*

Marcus Prior, *Post-Broadcast Democracy: How Media Choice Increases Inequality in Political Involvement and Polarizes Elections*

Paul M. Sniderman, Richard A. Brody, and Philip E. Tetlock, *Reasoning and Choice: Explorations in Political Psychology*

Karen Stenner, *The Authoritarian Dynamic*

Susan Welch, Timothy Bledsoe, Lee Sigelman, and Michael Combs, *Race and Place*

John Zaller, *The Nature and Origin of Mass Opinion*